To Richard.
Good Luck and Health
Many Thanks
10, John Davie

# Farming for Better Profitability

W. John Davies

Email : wdavies238@btinternet.com

ISBN : 978-0-9932541-0-9

First printed on 27/11/2014

Self published by W. John Davies

Printed in Wales by
PROPRINT, Carmarthen

# CONTENTS AT A GLANCE FOR QUICK REFERENCE

# Foreword

As I am now in my early eighties, I feel it may possibly be beneficial to recall some of my lifetime's experiences, and more especially my involvement with cattle and other farm animals, starting off as a farm worker.

The main object of my writing, is to try as simply as possible, to explain my personal experiences over the years, my understanding of cattle, and especially the soils they live and depend on, and how they are interlinked.

Everything I have written herein I believe to be true and genuine, written in good faith and honesty, as far as my memory allows me, regretfully, as said I will not see my eightieth birthday again, and I emphasise that I write as a hardly educated country bum-kin.

I have only recently become familiar with a Laptop and I assure the reader that I have not researched any material written here.

I admit that recently I did ask on google, "Why do cows and calves like to eat Newspaper." I did not avail myself of the answer offered on one website, simply because I decided not to pay the sum they asked for it.

As science progresses continually, the reader should never avoid or ignore modern Veterinary advice, simply because of what has been read within the following pages, I cannot accept any responsibility for any failures you may experience due to what you read, neither can the printers, or anyone else involved with this work.

However, I sincerely hope that you will enjoy and possibly benefit to some degree from what I offer you, and may you find some instances within, at least interesting enough to promote some debate, and hopefully at least inspire our  highly essential agricultural community, to stop and think a little occasionally.

I certainly do not claim that what I have written is all correct, it is up to the reader to say so if he/she thinks that I am all wrong, and I would dearly love to hear if that is the case, all that I have written about are my own  thoughts on my personal observations, today I occasionally allow myself to ponder, and I feel that I was ever so privileged to have always been so lucky to be interested and absorbed in whatever I have endeavoured to do, and that throughout my whole life span. Inherent interest in our work, I believe is the only main factor and difference between complete success, and failure.

Yours Faithfully,

John Davies.

P. S. Many thanks to members of my family for all the help, technical and otherwise that I have received whilst writing, and a special thanks to my dear wife for having such patience to put up with me during this time.

I also feel that I owe a deep debt of gratitude to my late grandfather who inspired me as a child to do my best at all times, he always emphasised that if I am rushing today, I have not got the time to do it properly, so  perhaps, it is best to leave it, and do it better  tomorrow!

And dear old Grandma also clipped me behind the ear once, I had said to her, "Sorry I can't", when she requested me to do something, I was told in no uncertain terms, "That three quarters of the word can't was Can." and that I better had not ever forget it, and I have not, to this day.

# Chapter One

## LEAVING HOME AND LIFE AS
## A FARMER'S BOY.

One month into my fifteenth year, I became a "farmer's boy" (a term that was used for a young farm servant). This early and important period of my working life started on the first Tuesday of January 1949.

It was a fine sunny and dry winter's day; at 4.00pm, after a nice tea, a highly qualified modern young Vet arrived on the yard to look at a young second calver Friesian cow that had given some concern for quite a while, she was by now too weak to even stand up, this Vet had attended this animal a day or two before and this time he reluctantly suggested that all we could do was to put her down. I personally remember his words clearly,

"Well, Mr Williams, I am very sorry, I don't know what is wrong with this cow, she is obviously suffering, and I am sorry we must have her put down." He spoke very quietly and sincerely, whilst he washed his hands before leaving. Saying "You better call the knacker man to collect her tomorrow, that will be the cheapest, and I'm afraid your only option in this case, sad as it is for you".

We eventually finished work at eight pm, on this my first day in farming, an introduction that made me realise at the very start what farming was all about, even so I was in a determined mood, knowing full well that I was doing what I wanted to do, I was there, fully committed in my own mind, to learn and to succeed as best I could.

Later that evening at 9.30pm, my boss suddenly jumped up from his chair and announced that he was not going to have this animal put down just now, well not until he had tried one old trick he had up his sleeve.

Next day he took a spade and an Enamel Pan, and dug out some fine soil from deep down in the middle of a wide hedge. When this was offered to the sick cow she looked at it, took one sniff, and promptly ate it all, licking the enamel bowl clean.

Having this soil twice daily for a fortnight or so the cow gradually recovered completely, she never once failed to lick the pan clean, and although still rather thin she calved down three months later, and again the next year, she was milking and still on the farm when I left nearly two years later in October 1950.

A farmer friend of mine had a similar case in the year 2005, some 55 years later, this cow was to be put down the next day as well, I told him the above story and he also achieved the same result exactly, by using soil dug from the centre of a wide hedge. Meeting him a few weeks later, I enquired regards the cow, and his reply was that it had worked for him equally as well. "I owe you a pint my good friend", he told me.

Another case I came upon quite recently, was an elderly farmer and his son who enjoyed showing fat bullocks each Christmas Show, the old boy did the feeding and he also included a good handful of dried hedgerow soil in each feed for his show animals, when asked about this, his words were "We will never win the Show without some good virgin soil in the feed." So there we go. Obviously, the Vets still did not understand this problem in 2005, as had been the case 56 years earlier.

On that same day, the first Tuesday of 1949 as said at 1.30pm, my new Boss had arrived at my home to collect me, and my total possessions, all contained in a tin trunk. He came in a very smart market trap drawn by a lively dapple grey Welsh Cob, so this was how I left the family home.

After just leaving home a very few yards, I was asked as we trotted along if I was able to drive the horse, and I felt proud and privileged to be able to do so, right on through town and further until we reached the farm, which was now to become my new home. Upon arrival, the horse was rubbed down and stabled, the trap was put away safely, and then my new Boss turned to me and said.

"We have not yet closed the main yard gate, you had better run up quickly, and close it, and while you are there please take a good look at it, and remember, should I ever in the future ask you to leave, due to any misbehaviour on your part, that you close it at that time as well, we always keep our main yard gate closed at all times".

So it was start off as you meant to go on. This was on a Tuesday, due to a strong local custom, that believed it was very unlucky to commence employing a new worker on a Monday, even so a full first weeks wage as it was at the time, would be paid.

In 1949 a full week meant a six am start each morning, and working on till six pm, each evening, Monday till Saturday, allowing just enough time to eat your food at mealtimes, with one Saturday afternoon off once only, each calendar month.

In addition I was also expected to assist with all essential farm work each Sunday, this meant, a full week of around eighty hours work, this was considered perfectly normal in 1949.

As well as the dapple grey Welsh Cob, there was also a darker grey Irish Draught horse, slightly heavier, this one had been working in a beer dray with a previous owner, he had not been broken in, or trained properly, and was therefore a little difficult to handle, when using long lines, as his mouth was so strong.,

This horse had a clock in his head, that told him the time of day to the minute, whenever the horses were working we usually stopped for the day around twenty to four, allowing me time to take the harness off, and settle the horses, so to be on time for tea at four pm prompt, just a few minutes sooner this horse would become restless, and would really let me know it was time to stop or finish work.

When living outside during the summer time, this horse was also very difficult to catch, that is except on a Friday morning, he knew he was in for a jaunt into town in the spring trap, as this was my boss's wife weekly shopping day.

This spring trap had a pair of shafts that extended each side of the horse, these were attached to the horse harness by short chains, towards the back end of the shafts and underneath them a generous dab of Stockholm Tar was always in place, a smelly, tarry and sticky substance, this was there to discourage flies during hot weather, as horses were very frightened of certain species of fly, especially the Red Robin that seemed to sting horses very hard, especially during very hot weather.

Daily from 4.15 till 6.00pm we were busy, looking after the stock of animals, and doing the daily afternoon milking that was still done by hand, if you were late finishing that was just your bad luck, the routine work simply had to be done.

Early this first spring I was told one sunny afternoon, to take the horses to be fitted with new metal shoes, my boss told me to be careful to wash their feet clean, and to keep them clean until I arrived at the Blacksmith Shop that did this work, if not he said, he will send you back to wash them again, simply as a practical joke,

On arrival this blacksmith looked us up and down, he knew the horses, but I was new to him, and as predicted, he promptly told me to take the horses home to wash their feet.

I was riding on the back of one horse, so I looked down at him, and quietly told him, that it would be cheaper for me to go on to the next village to get them shoe'd there.

"Oh!" he exclaimed, "In that case you better jump down and get them in here quickly". Somehow that instance seemed to give me a little more confidence in myself.

This first employment was on a rather hilly farm, the farmer was a very keen cattleman, in his small, but then usual way, and the holding of around fifty acres, kept 12 Friesian milking cows. There would be 6 or 8 younger animals kept for replacements.
Bull calves were sold at the local market at a fortnight old.

There was also one Ayrshire cow here as well, a very proud cow, with a magnificent three foot wide pair of horns, this animal developed a problem in one eye, and the local Vet turned up at six pm one evening to see to it, this cow was not having any of it, and used her horns expertly to tell us so.

My Boss had commenced farming himself in 1936, and during his first year he had invested well, when he bought two young Friesian heifer calves, these being the very best stock available at the time, from a fairly local top Friesian breeder, the cost being the princely sum of £4.50 each; a tremendous price compared to ordinary calves at fifty pence or so at the time, this proved to be a very shrewd move, and these were still on the farm 13 years later as elderly cows, but still milking when I arrived. Every one of the other Friesian cows, all various ages, were home bred out of these two original calves.

During the late 1940's cattle breeding improved immensely, due to the introduction of artificial insemination for cattle, this meant that for the very first time, there was access to better bred Bulls available to every farmer.

There was also another young cow, a large red cow, like a Red Devon, a completely different breed. This cow had been bought on a neighbour's retirement clearance sale for £12, which was considered to be quite expensive. When a local farmer retired, his neighbours would always support him, by helping him to get everything ready and ship shape, by the day of the sale,

Farmers also considered it to be very important to get their name on to the sale record book, simply by buying something or other, this was perhaps a welsh local traditional custom, maybe there was an element of satisfaction and pride in showing the local populace, that they could afford to spend a few pounds or pence extra!

6

This red cow never completely mixed with the Friesians, always a little aloof in her attitude, she would always be on her own, but yet, never far away, and she also never seemed to stop eating.

She was constantly browsing along the hedges, and would eat almost anything in comparison with the Friesians, who were certainly a little choosy in what they ate. Years later, this cows behaviour, helped to convince me that bought in animals never completely settle down and adjust to their new surroundings, so therefore they do not  go on, to produce of their very best, that is after being moved from one farm to another.

Government rules in 1949 meant that farmers of this acreage had to grow nearly one acre of potatoes, even though the Second World War had finished four years previously. Local farmers considered this to be totally unnecessary, and a waste of good grazing land, so a very steep south facing slope was cleared and prepared during early spring to grow these potatoes, the rows were opened a full yard apart, that is much wider than was necessary, simply to avoid some of the extra cost of seed.

That year turned out to be a dry glorious summer, and we had a phenomenal crop of beautiful potatoes; nobody could understand why that year was so exceptional, but sixty five years on, having in recent years been lucky to have the opportunity to grow top class potatoes for showing, I fully realise why this patch produced such an outstanding crop all that time ago.

The main reasons for this success were, firstly the fresh soil, secondly the south facing slope, and the extra spacing of the rows, these last two factors provided a lot of extra light to the growing crop, these good light conditions are essential if you want better potatoes than average, by now it is known that even a shadow caused by a tree, can reduce the crop considerably. We will go into growing potatoes in a little more depth in our last chapter.

One amusing incident perhaps worth mentioning was the fact that my boss was partial to a few extra cigarettes, over and above what his wife allowed him, he would often send me on my bike a good mile or so, to fetch an extra packet, this was usually just before tea time, and I was to hide them in a hole inside the stone built cowshed wall on my return.

On one occasion I got a puncture in the back wheel of my bike, so now walking, to avoid further damage to my tyre, I arrived back a little late for tea. Madam soon enquired why, and I saved the day's tranquillity by saying that the young calves had strayed out on to the road, and that I had fetched them back, the boss was so relieved that I had saved the situation.

Next day I had to take her Ladyship shopping in the spring trap, and whilst at tea afterwards she gave me a half-crown coin, (twelve and a half pence} for showing initiative by bringing the calves back on the previous day, this was a very nice and welcome surprise for me, and the boss could only glare at me across the tea table, and remain silent. That half-crown was easily enough to enable me to go to the pictures at least twice.

By December of that year, there were a goodly number of poultry fattened, and a week before Christmas, half a dozen local ladies turned up to feather them, a wooden frame had been set up to hang the birds up, so they could be slaughtered, the Boss realised that I was an old hand at this, so I had a very busy day catching these one by one, hanging them up, stunning and bleeding them by cutting a main vein in the back and top of their mouths, a good and complete bleed was very important, gradually I dispatched them all, turkeys, geese, ducks, and chickens, my grandpa's lessons in sharpening my penknife stood me in very good stead that day, and I was only two weeks into my sixteenth year.

During the winter the cows were fed mainly on loose hay, that had been harvested carefully the previous summer, just as it approached maturity. This hay was full of hay seeds that were then considered a very important part of their diet, all through the winter this was always ever so carefully rationed according to the size of each cow, with some extra corn and dairy cow cake for the ones that were milking.

Early 1949 saw cow cake price increased to £18 pounds a ton. The animals were all tied by the neck indoors all winter, but were turned out most afternoons for a walk on the yard, and were looked after extremely well. They would live outdoors from Mid-April to Mid-November.

On average, the total earned from milk on this farm at this time would be between £100 and £120 per month. This income equated to about ten pounds per month per cow. This was then a reasonable living. Milk was priced at 1/6 (seven and a half pence) a gallon (4.2 litres). Bearing in mind that there were 240 old pence to the pound in the old imperial money, this works out at roughly five gallons of milk per cow a day, which was pretty good going.

My wage this first year was £5-00 a month, with 3s.9p (18p) deducted for national health stamp each week. This left me with £4.09 pence per month in my pocket, plus free board and lodge.

Throughout the whole year there would be a five gallon drum half full of diluted disinfectant somewhere in the cowshed. This had a tiny hole near the bottom that allowed the contents to seep out slowly onto the floor, this drum was moved around daily, from one cow stall to the next, the result being a strong prevailing smell of disinfectant around and above the cattle, and also the whole building. The word 'above' in the last sentence is I feel very important even for modern farming today.

It was not until I had retired fifty years later, that I realised the full advantage of this, as it certainly confuses air-borne bugs and flies, that are definitely attracted by natural animal smells, these bugs bring in new diseases with them, by now the new diseases such as Blue Tongue and Smallenburgh, springs to mind, deodorants for animals may soon have the potential of avoiding much stress for farmers and their animals, that could possibly increase profits very substantially, again we will certainly discuss this in greater depth a little further on.

The next winter we trapped the rabbits on the whole farm, they were endemic at the time, a dry hilly farm was very attractive to rabbits in the wet winter months, for the first few mornings we went around the fields to collect the rabbits caught, they became less in number each day, then on the fifth day the Boss told the farm dog, Rover, to go around the traps by himself, this was just as we were going into breakfast, when we came out Rover was not there as usual, so I was told to go round the traps alone, and sure enough I found Rover lying down five yards out from the hedge with a rabbit trapped right there.

This dog had been trained as a youngster, never to go near the hedge when there were traps put down, had there been no rabbits caught he would have been back home when we came out of breakfast, apparently he had done this for several years, which made the patience needed to train him very well worthwhile.

My understanding is this, you spend ten minutes a day for two or three months, to train a ten month old dog fairly thoroughly, then, with a little luck, you have a well-trained dog for ten years, this is surely a good return for using a little patience, ten minutes a day is just about as long as a young puppy can concentrate on any one thing, at the beginning of training.

The main important point is the handlers ability to get the dog to listen and stop, like riding a motorbike, you are in trouble very soon if you cannot stop, another very good point when training a dog is never allow him to jump over a wire fence or a gate, this is not a good habit and inevitably, as the dog gets older he will catch and break his back leg, causing him much sufferng and this will cut short his working days, by as much as two years, when this happens it is a great loss of an experienced and faithful worker, leavng a tremendous gap in a farms work force, especially if there is no working dog ready trained to replace him. A dog jumping a fence is simply rushing,  like us humans a dog needs and deserves a little time, to do his job properly and also safely.

On average mid April was the time to turn the cattle out, so they could live outdoors for the rest of the summer, the evenings were getting lighter, and so my boss would lead out the new young cot calves one by one, and naturally they would struggle vigorously, as they had not been tied and handled before, but when they cooled down a bit, he would lead them up to a barbed wired fence, and would calmly push their noses up against the wire.

The spikes would cut their noses rather badly and they would bleed quite strongly. "There you are", my Boss would say. "If you touch that fence it hurts you", exactly as if they understood him, being all home bred every animal whatever their age  had this same scar.

Again this made me realize much later on in life, that young calves, (and also all other young animals as well) never forget any incident that happens to them, even from day one of their lives.

Understanding this, and the reason why the red cow always remained aloof of the other cows, will result in better all-round profits eventually for all stock keepers, we will also discuss how to turn this to advantage later on.

Four acres of Black Supreme oats were grown annually, this would be fed as whole crop to the cows each winter. When fully ripe this variety would be standing three feet high at least, we avoided trouble and losses caused by marauding crows, by making sure that our neighbours, had sown their corn at least a week before we did in the spring; their corn would then ripen before ours, and would attract the crows attention away from our field, that would normally become fully ripe and ready to cut during mid to late August, the contractor had agreed to come in two days time to cut the crop.

The field had now to be opened up, that is the corn around the hedge had to be cut by scythe, so that the tractor did not trample and waste any as it went around the field perimeter for the first time.

Remembering my grandfather's thorough instructions, I started to scythe around the field cutting a full eight foot wide swathe, and thinking nothing of it, but my Boss was so impressed by what he saw, that unknown to me, he immediately telephoned to stop the contractor coming, so in two and a half days of glorious weather, I scythed the whole field of corn completely by hand, that is between the morning yard work, and the evening milking, this was well equal to the olden days, when a man was expected to mow one acre a day of hay, and one and a half acres of corn. This crop was turned each day by hand using pickforks only, so not to lose the grains, and it was well dried and safely in the barn on the fifth day, using fully the spell of good weather.

With the harvest over, the next job was to trim all the hedges, this was also all done by hand, starting on the roadside, it was emphasised on me to do a good tidy job, as there were a lot of unofficial inspectors about.

As a raw fifteen year old it took me a few days to realise what this meant, and I never once saw an inspector, with an official badge, but there were plenty of local passers-by, either walking or on bicycles ready to pass remarks, if my work was a bit untidy, if nothing was said my work was OK.

In September it became obvious that there was a surplus of home cured bacon, left over from last year. I offered to buy one Ham for the asking price of £7.10s (£7.50). Looking at me suspiciously, my Boss inquired "How can you pay for it?"

I promised to pay promptly the next morning. The Ham was duly packed in a hessian sack, and tied to my bicycle handlebars. There were a few pubs in town, and I sold the Ham with no trouble for £14.00, I personally, had never handled so much money ever before, this of course was illegal black market trading in those days, getting caught could have meant a custodial sentence.

The local police were always on the lookout and not completely innocent, had they spotted me they would have probably offered to let me off, so long as they had half the Ham for themselves, but I got away with it.

The next morning I paid up, and my Boss looked quite puzzled as to how I had managed it so quickly, however, three weeks later, whilst having a drink in another pub, he found out what had happened in full detail, including the price I had been paid.

He was not a happy bunny for some days. I told him that I had paid his asking price, and the profit I made was a bonus for me for cutting his corn, after all I had saved him the expense of the contractor.

My profit, was more than equal to six weeks wages, and was very useful to pay off my debt on my bike.

My yearly agreement to work was to end on the 12th October, as was the custom. So during September, some hard haggling went on, and eventually I agreed to work for another year for £2 a week, plus my keep, payment monthly, same terms and conditions as before.

The work was now more or less a repeat of last year; looking after the cattle and any farm maintenance needed, including the annual ritual each spring of white liming all the buildings around the yard, as well as the three storey farmhouse, there always seemed to be a list of jobs waiting to be done.

During the winter we carried water to the cows in buckets, except on a fine day when they would be turned out to drink from the pond nearby. They were then allowed to stay outside for a few hours. This obviously rested their feet from the concrete inside, and I can honestly say that we never had any foot trouble the whole time I was there, but to be fair they were always well bedded down with straw.

The range, or passage in front of the cows known locally as the walk-way, was used to store hay brought in from the loose hay barn, this walk-way was allowed to become empty by the first Sunday, of each month throughout the winter, on this day the animals had no food at all, only water, the amount of milk sold was slightly down on the Monday after, but this was considered very worthwhile, my boss believed strongly that the cows benefited greatly from this one day fast.

The range would be swept squeaky clean on this day, and the considerable amount of ripe seeds that had fallen out of the dry hay, would be collected up, and this would be mixed up daily in the cows feed, the extra oil in this seed, would keep the cows coats shining like glass, and if there was a mouse around long enough, she would be shining like glass as well, as mice really love eating grass seeds.

But one bag full of this seed was kept on one side, always hanging high off a beam to keep it away from mice, this was to repair any damage, that had been done to the grassland, while working the horses during the wet winter, bearing in mind there were only wooden wheels with metal bands on the rims, these would cut the land up rather badly, carts would only be used on grassland during periods of hard frost. So in the early spring of the year, this grass seed was sowed on top of the soil where ever it was needed and just left on the surface this was known as Patching, this never failed to germinate well, and covered up all the damage, that had been done previously.

There was another incident that happened during the summer of 1950, that was also a mystery to me at the time. One afternoon the cows were patiently waiting on the yard, to be let in for milking. Suddenly, two goats appeared from nowhere.

The cows were completely spooked by the goats, and so they bolted off at headlong speed, with their tails up in the air, down the fields, all bar the red cow that calmly followed them at walking pace. They were found cowering and actually shivering under some over hanging trees in the boundary stream, at the bottom end of the farm.

It took me a whole hour to persuade them to return to the yard. In fact, the red cow that remained calm and collected actually helped me to convince them that everything was back to normal. This unforeseen behaviour was a complete mystery to me at the time.

Twenty years later on as a nurseryman I constantly had to adjust the growing medium, or soil and composts we used, so that they suited each individual variety of plant or shrub that we were continually growing to sell later on in their season, and lime was found to be the most important element that always had to be adjusted, so to be bang on, at the correct level to suit each variety, get it right we could achieve equal success across the board, and I gradually realised that too much lime on grassland, could hardly ever be a good thing, especially for the millions of natural microbes and bacteria in the soil, and likewise for the animals that graze on it.

I believe very strongly that natural Trace Elements are essential to all animals, at all times, the cows panicked probably due to a lack of one or more of these elements.

The sick cow was short of the same elements also, and the farmer found these without realising it, deep within the hedgerow, where no lime had ever penetrated.

Up to the early forties, which is just before the time we are talking about, very little artificial fertilizers had been used, on farmland, then during the second world war years, and again up till 1948, artificial fertilizer was only used very sparingly, compared with present day practice, but regular use of lime was thought to be good farming practice, the burnt lime used was very strong, and mostly spread by shovel very haphazardly, and certainly the amounts used were over generous, for the good of the animals at the time, and almost certainly this lime was either locking in or neutralising the most essential Trace Elements that were present in the soil, and thereby depriving the animals of their benefits, the results being highly nervous animals.

We will discuss and try to understand the importance of getting these Trace Elements correct in our soils as we go on quite a bit later.

Another detail on this farm was a fence five yards away from one boundary hedge. This was to keep the cows well clear of the neighbouring farm, that also ran a cattle transport business.

In those days the cattle lorries were cleaned out at home, thus bringing home all possible health problems. Our two working horses only were allowed into this fenced off area. This emphasises that the cows were highly valued and looked after, as well they should be.

As the summer of 1950 progressed, it became clear to me, that my boss, would look for a younger and cheaper servant for the following year.

This meant the end of my short relationship with the two working horses, this fact left me rather sad, for quite a long period, after all, we had developed a solid trust for each other, but I was ready to move on to gain more experience.

The afore mentioned pond was a very useful asset, being on the farm yard, each year it would fill up somewhat with soil and sediment, and although it was nearly on the top of a ridge, water would pour in and out of it during wet weather, so once a year on a particularly wet day, one of the horses would be brought out, and I would jump on to a thick hessian sack on his back. I would also have three or four sacks draped all over me, the object was to keep the horse moving smartly round and round within the pond so that his feet would disturb all the muck, this was taken away by the overflow, I do not recall any concern about pollution, but this was a quick and efficient method of cleaning out the pond.

As I was leaving this farm for the last time, a neighbour called to me as I went passed on my bicycle, he wished me well in my new job and said "I hope you realise that you have had two years of the best college available in this country today, and I hope that you go on to make the most of it." Truly the older I get the more I agree with what that kind gentleman pointed out to me that day.

# CHAPTER TWO

Life on my Second Farm,
Getting used to Tractor work,
Handling the Bull,
The experience of Brucellosis.


So on the 12th of October 1950, now coming up to 17 years of age; I started working on my second farm, this time of 140 acres.

My agreed wage was £3.50 a week, plus board and lodgings, with every other Saturday afternoon off. (Working conditions were slowly improving) this was then a very high wage, for a lad of my age.

I think it was my ability to use the scythe that justified this, and I soon found out, this farm had six acres of kale and rape as green crop, which had to be cut and brought in fresh for the cows, each and every day throughout the winter.

There were around 30 milking cows, milked by machine, a new experience for me, with around 80 animals on the farm altogether, there were two basic Tractors fuelled by petrol, which was used to warm the engines up, then once warm they could be switched over to Tractor Vapourizing Oil, known as (TVO) this was cheaper, but we had to be careful at all times to keep the engines quite warm, should they be left just to tick over for a short while during cold weather, the engines would cool down and they would soon start miss firing, the ignition plugs would oil up very quickly, to rectify matters we would then switch them back onto petrol until they warmed up again. Most tractors of that era had a canvas apron that could be adjusted up and down in front of the cooling radiator, this restricted the flow of air, and was slid up and down according to the weather.

All the implements used were trailed behind these two tractors, everything mechanical was completely new to me, so I was now on a steep learning curve again, on the farmyard itself the main tools used at this time were a hay knife, to cut all the loose hay out of the haysheds, also used later to cut out silage from the pit, a couple of hand forks and shovels, a wheelbarrow, and a hard cane brush, there was a corn crusher in the barn, used to crush the home grown cereals each winter, a saw bench used to saw firewood for the house, and occasionally we would saw out a plank or two and some wooden stakes, on one occasion we broke the wooden pole that hitched the hay trailer onto the tractor, and it did not take very long to saw a new one out of a tree trunk and fit it back on to the trailer, and off we went again.

This farmer quite rightly expected a full day's work, but was rather less conscious of details, compared to my previous boss, especially when it came to the cattle. I soon adjusted and settled down to his ways, and soon realized, that there is always more than one way to do most things.

His pride and joy was a young Friesian stock bull that he had bought rather expensively the previous year for the sum of £110, he was rather well bred, and he grew to a massive size.

This bull was always tied at the bottom of the cowshed, in a double stall, shared by a small grey cow beside him. He soon became difficult to handle and quite ferocious, in fact positively dangerous.

It was part of my job to handle this beast each time his services were needed, he was always led by a chain that was linked to a ring in his nose, with this he could easily lift me bodily off the ground. All I could do was turn him around and around, to control him.

I soon benefited from a story of a local farmer, whose two bulls had also become uncontrollable, these were sold for slaughter, the following day a lorry arrived to take both animals away, whilst the farmer was having his dinner. The driver enquired of their whereabouts, and the farmer told him, "I will get my two assistants to help you, as these bulls can be dangerous."

The driver said that he did not need any help, and promptly to the farmer's amazement, led the two bulls into the lorry, one by one having no trouble at all, but whilst cleaning out the stalls later on, this farmer found two complete eyes on the floor!

That driver had callously scooped out one eye for each bull. It was no wonder that they went quietly!

Knowing this, I found that a gentle slap across the bull's eye, his only weak spot by the way, was enough to keep him under control without any damage or cruelty.

As he became older this bull proved rather more difficult and there was talk of getting rid of him, but surprisingly he was fairly quiet within a week or so, and remained so for nearly two years, before he was sold on.

I never shared this my secret with that farmer, and he never fathomed out what I did either, although he was extremely pleased with his now quiet bull.

When he was eventually sold and taken away, I was away on my annual week's holidays, and the bull was once again starting to create havoc, as he had been when I had my previous year's holiday.

The new purchaser didn't mind his aggressiveness as he had a new impressive bull pen to handle him without danger. However, within a month the bull had been shot for being completely out of control.

All that I had done to keep him quiet, was to place a pebble under the tongue of his drinking bowl, he could now only drink when I decided to remove the pebble and let him have one.

Nobody spotted what I did, but while slightly thirsty the bull was rather better behaved! Considering the ease with which he could be controlled, it was a shame that the new owner shot him. At 79 years of age, 60 years ahead, the behaviour of that bull is by now quite understandable to me.

Bulls are naturally inquisitive; they like to see what's going on around them. Keep a bull out of the daily proceedings of a working farm, and especially in a dark shed, he will very quickly become bored, and consequently nasty and impatient. Keep bulls where they can see, and hear everything that is going on, they usually remain quiet and docile.

However, there is one plant if fed to bulls, will soon make them very aggressive, and that is Horseradish. This is a herb and a member of the Mustard family, it looks like a strong Dock leaf but with a twist in the slightly longer leaf.

Our Horseradish sauce is made out of its roots, once established and growing this plant is virtually impossible to get rid of, as the roots go very deep into the ground, leave a tiny bit of root, it will always grow again, it is easily identified by taste, the leaf has rather a strong potency, and more so after being dried while making hay, as is the case with all dried herbs, they are much stronger than when cut fresh.

This Horse radish has usually established itself in grassland after seed has escaped from neglected gardens nearby. I have also known one instance when the quietest of Hereford bulls, suddenly went berserk due to the presence of this plant, this was on a farm that used to have a small mansion nearby, no doubt the gardens there had been rather important a few years earlier.

During my second year on this farm and now earning a wage of £4.50 per week, I had the awful experience of Brucellosis. Not much was known at the time about this.

The first we knew during early spring, a cow had aborted her calf, at around half the gestation period, soon there were more, one morning we found four calves had been aborted during the previous night: a truly heart-breaking experience

The Vets recommended getting rid of the cows, but the decision was made to keep them on, and get them cleansed out internally, and eventually they all became all in calf again, naturally the milk yield was badly affected for that year. This Brucella was an infectious disease, it left me personally, with a chronic tight chest, and I was continually thereafter in a mild cold sweat.

At the time I was told that this problem would wear off if I ever saw my sixties, and my health did improve between the ages of 60-65, true to the doctor's words.

All milk was treated and pasteurised at the factories of course. It was the fresh milk we drank at home, that caused the problem.

During my time on this farm I experienced some very difficult calvings, and while assisting the Vets I learnt quite a lot. We would still sometimes loose a calf, but most of the difficult births were eventually successful. I was present and helped with possibly the first Caesarean operation known in this area, which was nothing short of a marvel, and a new experience previously unknown to us, all operations were usually carried out by the younger Vet in the local practice.

Two operations also took place on cows, with wire, in their stomachs; these were both successfully performed on Sunday afternoons as they took rather more time to complete.

One calving sticks out in my mind. The small grey cow that was always kept beside the bull, was due to calve. The calving proved to be a particularly difficult one, after much effort the calf was delivered alive.

The cow was turned loose so that she could lick the calf dry, as nature intended, immediately she charged at me, putting my hand out to defend myself, I found I was able to push her headlong and sprawling onto the floor, without any pressure at all. It was obvious that her brain was not fully in control of her actions.

Thinking about this behaviour years later, the bull must have been regularly, eating most of her food ration. This part starvation, and the calving effort, tipped her over the edge, but with a little careful feeding she very soon recovered to her normal mind-set.

Again several years later I came across a similar case with a bunch of large bullocks, they were literally starved, by being left out in a wood to fend for themselves, all winter. These animals would also charge murderously at anything and anybody that moved, but were too weak to be of any danger to anyone. Their minds and their limbs were completely out of co-ordination due to wanton neglect and semi-starvation.

There may seem to be a fair bit of grass and various greenery in woodlands, especially if the trees are rather sparse, but due to different species of michro-rhizas in the woodland soil, the grass growing there is not suitable to sustain animals for any length of time, they will survive for a period, but they will hardly ever thrive, especially if they are given no supplementary feeding.


## Chapter 2.2

Advances in technology. Silage in the early days,
First de horning of Cattle, Eradication of the Warble fly, Stopping Ringworm for pence.

During my three years on this farm (1950-1953) farming advanced considerably. Autumn /51 the electric wire fence arrived for the first time.

Having never been seen before, this was great, as it controlled livestock so they could graze green crops outside during the winter, one result being there was no more need for my blessed scythe! except **that** I used a short bladed one when trimming grassy and ferny hedgerow tops, this was so only until two years later, when the first mechanical engine and belt driven hedge cutter, all mounted on a trailer, arrived on the scene summer 1953.

Also during this period the first diesel-fuelled tractor, a Fordson Major was purchased, this now had the first hydraulic arms, fitted at the rear, that could lift, and carry loads of around half a ton of weight, the buck rake also came on the scene, again to use on the back of tractors.

This was a blessing, we could now carry fencing materials around the farm, this saved dragging a trailer, and making a mess wherever we went when the ground conditions were too wet, the buck rake also coincided with the actual beginning of making silage, in West Wales, initally in small quantities as a trial and error, by using a mounted buck rake, we could harvest about fifteen acres of silage a week.

This seemed the best way forward, and soon developed rapidly. Making silage could carry on when the weather was not quite good enough for hay making, this was now cutting out the risk of losing hay-crops, due to bad weather.

Pits were dug out of banks so that silage could be stored underground, when eventually filled after several days of collecting the grass off the fields with the buck rake, the Silage pit would be sealed with approximately twelve inches of soil, and sometimes we used Lime on top of this, simply to try keeping the crop airtight, as polythene sheeting had not yet arrived on the scene.

During 1951 we also started cutting the horns off the milking cows. Previous to this, dominant cows could do considerable damage to younger animals with their horns.

Under the Vets supervision the cows head would be injected with anaesthetic to kill any pain, some twenty minutes before a rope was pulled as tightly as possible around the base of both horns, the cow was now held as steady as possible, and the horns would be sawn off with a metal hacksaw, inevitably, numerous blades were broken, and there would be blood spouting everywhere.

This blood coagulated, eventually, all over the place as well as on the animals face. Thankfully, we soon progressed to using wire, just like cutting cheese.

After a few mistakes, we noticed we could see the veins pumping the blood out, we soon found we could stop the blood flow quickly with tweezers, and we could pull the bleeding veins out, and tie a knot in them. Soon we moved on to using a hot iron to sear the veins, and thereby easily stopping the bleeding.

Sometime later, de-horning of young calves at ten days of age became the norm. A hot iron designed for the job would remove the young horn bud; this was a great improvement, that saved a lot of time and effort.

Again, during the early fifties, an eradication programme to get rid of the Warble fly became law. This meant that all cattle, during late winter had to have their backs scrubbed with a strong stiff brush, and then treated with thick liquidized Derris powder , this was done to break the life cycle of this dreaded fly.

Apparently, this fly would lay its eggs, on the hairy legs of cattle during the hot summer days, this was very painful, and the animals were terrified when they heard this fly buzzing above their heads, before striking just above the cows feet, when hatched the larvae would burrow under the skin, and migrate, until they reached their goal under the hides, on the cow's back, eventually emerging again as large lumps in early spring, leaving large holes that often became infected, these holes left the skins on carcasses valueless.

Over a few years this treatment worked well, and eventually got rid of this awful pest, that has thankfully remained more or less extinct ever since. This improved the Carcass value of animals, for a short time, due to the better skin values, but I feel this aspect was very soon forgotten about.

During June 1952 a spell of dry weather meant that all hands were turned towards hay making, for me personally it was a 3.30am start each and every morning, I was to mow, as much hay as possible before breakfast at 9 am each morning, the rest of the day was soon spent turning and drying the hay, each field at a different stage.

One late afternoon a thunder storm suddenly threatened. We had nine acres of prime hay ready to be brought in to the barns, we started carting around 7.00pm hoping to get a load or two before the imminent rain started, thunder and lightning flashed around us but no rain, it turned out to be a rare dry storm, and we completed the carting at 2.30am the next morning, an hour or so before the rain eventually moved in, the hay came in perfectly dry, as there is never any damp dew whenever there is heavy thunder about.

Next day it rained all day, every one of the family went missing, and I slept all day on top of the hay shed, nobody missed me, after all I had completed a twenty-three hour shift the day before.

One summers afternoon, on a very rare occasion, I was left by myself to do the evening milking, this in itself was unusual, and when I prepared to milk one animal, to my horror I realised that one of her four milking teats, was not there, it had somehow been completely torn off cleanly, with only the milk duct hanging loosely, there was no sign of any blood.

All I could do was phone the Vet and ask his advice, he confirmed, that nothing could be done, but that I should smother it well with Stockholm Tar, a black substance with a very strong smell to keep the flies off, as was done earlier under the shafts of the market trap (chapter one), confirming once again that flies are attracted to their prey, by using their keen sense of smell, the significance of this observation is very important, and will become more so, if our climate tends to warm up as is being forecast, I did not realise the importance of this, for many years after, and we will again give this more important attention later.

Another incident that happened while on this farm is also well worth mentioning. One winter we had an outbreak of Ringworm, affecting the young stock, and inevitably, I soon had a large patch of it on my left wrist. This was very irritating and Doctors gave you some purple stuff to treat it, all to no avail.

On one very cold February morning the tractor I was using started playing up, while tinkering with the fuel system I somehow got some petrol over this infected wrist, within a few days my wrist started healing up well.

I soon found out that Diesel oil was equally effective, in clearing up this nasty little parasite, apparently this bug lives on, or in, the dust that settles on timber, in the cattle sheds, so a simple answer is to paint or spray all timber inside cattle sheds with a mixture of diesel and soapy water, I never since had any more ringworm after doing this regularly each spring after turning the cattle out.

Also each winter regularly each month, I went round all my animals lightly rubbing their coats all over with a rag, again kept moist with paraffin and soapy water, a half and half mixture, and paying a little more attention around the base of their ears and around the rump and tail areas, this is effective enough to keep all manner of lice at bay, this also kept the animals looking so much better at turnout in the spring,once again for hardly no cost at all.

In addition to the milking cows on this farm, we also produced 30 acres of Cereals as well as green and root crops, this along with an acre or so of potatoes meant that the farm was mostly self-supporting.

Each spring the hayfields would be given a dressing of fertillizer as soon as the weather started to warm up, then in May the hay fields would receive another dressing of Nitro chalk, even though the crop would be already around eighteen inches high, what the tractor wheels rolled flat soon recovered, and the crop greened up very quickly, this was done about three weeks before we expected to mow for hay, however if the weather turned wet for some time  and the mowing was delayed, we would then have a devil of a job mowing it, as we only had trailing fingerbar mowers at the time.

As said this Nitro chalk a straight nitrogen would cause the loose harvested hay to warm or even heat up considerably.

Soon after carting into the hay barns, long steel bars would be pushed into each bay of the hayshed and checked daily, if these became too hot we would then sink a two foot square hole in the centre of each bay from the top of the shed down to the floor, this was hard warm work to say the least, we would go down the hole using the hay knife to cut as deep as possible, then gather the cut hay in our arms to take it back up to the top with you, this went on till we reached the bottom, on one occasion this hay was so hot that it burst into flames as it had extra air when I got to the top, this was really too close for comfort, but we managed to save the situation as the rick gradually cooled down over a period of a few days.

My boss was exceptionally keen on getting good dry well harvested hay, after mowing he always concentrated on drying the ground first. He always said, "Dry the ground young man, and the hay then will dry itself," This was achieved  by raking up the mown hay as soon as possible, then moving the rows again before spreading, this method stood me in good stead thereafter all my life, and my hay was also equally ready a good day sooner than the norm.

Each spring on this farm there was a surplus of hay that was sold off to various farmers; this was loose hay then of course, as the Baler took another year or two to arrive.

The experience of growing these crops using very basic implements, all towed by tractors, proved very useful to me, simply experiencing this farmers determination, to get things done against all odds, stood me in really good stead when I became self-employed, on my own little nursery a few years later on.

While on this farm I had the experience of seeing the very first foreigner I had ever met or seen, that is barring the American soldiers and the European prisoners of war I got used to during the War years, it was the summer of 1951, a dapper little fellow turned up on the yard carrying a suit case, he politely explained that he was from Pakistan, that he now lived in Swansea, and that he had come down to our district by train that morning, to try and sell some Silk Ties that he had in his case, he had a seven mile route mapped out which he walked calling in each house and farm as he went by, his aim was to complete his route and catch a train back home later that afternoon or evening, eventually off he went on his way, a month or two later he turned up again, this time with a variety of ties and scarves etc., he insisted that I buy a scarf with him, so that all the local girls would be guaranteed to love me so much more, looking at his watch he suddenly became a little agitated and he asked what was the time, it turned out that he only had half an hour or so to catch his train, I pointed out to him that there was a footpath he could use to save going all around by road.

I was going along this path to repair a hole in a hedge, so I collected my tools, an axe, a billhook and a spade to do the job, but he was not happy to come with me. After explaining to him, he then followed me at a distance obviously very wary of me and my tools. When I arrived at the hole in the hedge and started working he came past me all smiles and off he went, in a minute or so he came back, and thanked me for being his helpful friend.

He soon turned up again, this time in an Austin ten van and he kindly gave me a silk tie for helping him out the time before, having started probably from nothing, these people made an effort to improve themselves, today we know that these people are almost running most of the retail businesses in our country, there must be a lesson here somewhere that we affluent British have missed out completely on.

Whilst on this farm it was decided to sink a new well, to hopefully supply the dairy with better water, this was before any hydraulic machines were being used, so the well was dug completely by manual labour, a windlass was set up, and all the spoils were wound up in a pail to the surface all by hand, after the first days dig we were lucky if we could go down more than twelve inches a day, water was found at twenty feet down, and all was hunky-dory and successful.

After the initial failure of the water to pass the necessary health tests, a spoonful of Deosan, a sterilizing liquid was added, and on the next test it passed no problem. After three years of good hard training, I decided to move on to another farm once again.

## Chapter Three .
**Learn**ing some more on my third farm.
Catching the family pony.
Myximatosis in rabbits.


Soon I had agreed to work on a dairy farm of 110 acres this time, again not far away, this was now Autumn 1953 and my weekly wage was upped to £5 a week plus my board and lodgings. Work, as before was a 12-hour day for 5 days, plus Saturday mornings (every Saturday afternoon off by now) and again, all essential work on Sundays.

This young farmer had established his milking herd, by buying fresh first calf heifers the previous year. With very good management they had milked exceptionally well the previous and first year. I was now fresh on the scene to take them through their second year.

The yields of milk this year fell well below expectations, in comparison to the previous year's yields. My new boss started to question my ability as a herdsman, but as I was following his instructions to the letter, I knew there was no fault on my behalf.

So it was soon amicably agreed that he would take over milking, for a month, to show me how it was done. However, the yields did not improve, and I was soon asked to resume the milking duties.

These young animals had produced their first calf at two years old, and then having milked heavily that first year, then back again in calf as soon as possible, the strain had taken its toll, milk production suffered that second year, bearing in mind these cows were still young and growing. In other words, nature was looking after itself. Even so, when on their third lactation the following year, milk yields improved once more.

This proves that a slightly longer period between the first and second calving gives young cows a better chance to grow to their full potential, so that eventually when they get to calving for the fourth time, they are at their prime best, and therefore of more value should there be a surplus of animals to sell.

Milking machines at this time were rather basic, the cows were milked into four gallon buckets and these were carried into the dairy, and the milk was tipped into a fairly high container and allowed to run down over a cooler plate that had cold running water running through it, later this summer we experienced problems with our milk, the routine tests by the dairies were showing that the keeping quality was not up to standard, despite thorough cleaning and maintenance of all the milking equipment on my part.

This came to a head when a milk inspector came along to try and solve the problem, which turned out to be that the milking liners had started to perish on the inside, and needed to be changed more often than had been done previously, this was the first time that we had come up against this sort of thing, the tubes and liners were now being made to a cheaper standard and were not lasting so long, at least I was left off the hook, as it was  proved yet again that it was not my fault.

During this period, (1953-1954) hardly any farmer that I knew, had any cattle handling facilities worth mentioning on their farms. To dose the cattle for stomach worms, and the annual compulsory testing by ministry vets for TB and so on, was rather a spectacle. The young cattle had to be rugby tackled and actually held by force. It could take all afternoon to catch twenty or so cattle.

How we avoided serious injury, was only a matter of luck. Thankfully the milking cows were mostly tied by the neck and were used to being handled. Being nervous of strangers, cattle always knew that something was up, when the vets arrived.

A strange voice, and the smell of a different disinfectant, was enough to put them on their guard, and they would become rather wary and nervous.

Most certainly, looking back now, almost sixty years, this nervousness in cattle was never even considered, to have anything to do with the lack of Trace Elements in their diets, that is known to be a strong factor today.

The average price for milking heifers in 1953-4, was around £40-£60. Occasionally a pedigree heifer was purchased for around £100 or so. These were given a little better treatment, and turned out to be a little more profitable even then.

Whilst on this farm during the summer of /54, we became aware of myximatosis, known as myxo, this was an awful disease that wiped out the whole population of our rabbits, within a few months or so they were all dead, mostly underground and in the hedges.

This also put paid to the rabbit catchers one and all, the following year without exception farmers were amazed at the extra amount of hay and silage they were able to harvest, the rabbits had been eating a tremendous amount of grass, and spoiling much more by urinating on it, previous to this disease many farmers could pay most of their rent with the rabbit money.

Speaking of rabbit catchers, these people during the period of recession before the Second World War were probably the richest people in our rural community, so much so, they used to advance small sums of money to farmers quite a few months before they intended to set their traps on their individual farms. The rabbit dealers were also very affluent, as they would have numerous trappers working for them exclusively.

Some trappers that considered themselves a little elitist, would only trap a farm once every five years, they would then tell a farmer to get, perhaps a part-time trapper to catch the rabbits for him in the intervening years, happy to let the numbers of rabbits build up again before they returned to cream them off once again.

I do not think that anybody, connected with agriculture fully realised how much damage rabbits were actually doing to grass pastures, let alone cereal crops and private gardens, in churchyards it had been fashionable to plant saxifrage plants on graves, this is a low spreading pretty spring flowering plant, its foliage was useful to suppress weeds, but the rabbits polished off all the flowering heads before they had a chance to flower and seed themselves, non-gardeners were amazed to see them flowering so proficiently for the first time ever, that first spring, after the demise of the rabbits.

One day late spring my boss came home with a nice looking welsh mountain pony, suitable for his young daughter, it was kept for a few days on the lawns around the house, and he seemed quiet and bomb proof with no vices, everybody was pleased with him, soon he was turned out onto a three acre field behind the house, so to be near at hand.

The following Sunday afternoon I had gone to bed for a few hours up till teatime, after which we would start work again.

Some relatives came to visit, and naturally went out so that the young girl could have her first ride, but the pony thought otherwise, and resisted all their efforts, there were now four adults and two children chasing and trying to catch the pony but to no avail, there was quite a bit of shouting going on and the pony worked himself into a lather, seemingly used to this type of a palaver, while at the teatable a little later the pony was going to be sold forthwith.

I enquired what was going on, and ventured to suggest, that they did not know how to catch a pony!, and that I would have to show them how it was done.  If I did, the boss was not very complimentary, soon a wager was agreed that if I could catch him within twenty minutes he would pay me an extra weeks wages, cash in hand, if I failed I was to forfeit a weeks wages, I could use whatever I needed to do so, and I could ask for assistance if needed.

I put on a large old overcoat, and collected a large armful of loose hay and went out to the field, walked around the pony at some distance, and then went off to the farthest corner where I shook the hay into a large loose bundle, hiding what I was doing by holding my overcoat wide open, I then walked away, and sat in the hedge not even looking at the pony,  but being curious he soon moved up to the hay, I then drove him off and gathered the hay up carefully. I then repeated what I had done twice more, never allowing the pony to touch the hay at all, by now the pony was more interested in the hay than anything else.

I then brought the hay near the entrance to the field and stood nearby, the pony came up  and started eating, I walked quietly up to him and held him by his mane, no problem, this took eighteen minutes to accomplish.

Very little was said at the time, and fair play I was given my fiver. This turned out good value for money for the family, after doing this a few times more the pony became easy to catch whenever they wanted to.

Years previously as a child I had learned this method by watching an old gipsy, that used to camp once or twice a year near our home, he was a friendly old soul and we kids enjoyed seeing him and his family turning up as they did. We spent hours watching and learning from them, and had many a row when we came back home, not realising that our clothes were reeking of wood smoke.

My time on this farm was spent mostly looking after the livestock, with most of any spare time spent trimming and laying hedges on the farm. This hedge work started very early in September 1954 and went on throughout the winter month's till mid-April 1955, by that time we had finished burning up all the surplus growth and trash that had been cut out, all this work was done with simple hand tools only, the hedges without exception grew well afterwards, despite the early seasonal start and the late finish.

That year on the last Saturday of November 1954, our country was subjected to one of the worst hurricane winds I have ever experienced, before or after, during that night a large number of trees were blown down on this farm, and also all over Pembrokeshire as well, and it took us practically a whole year to get them cleared up, we only had cross cut saws, trimming hooks and  axes, to deal with them, mechanical chainsaws only became available a year or two later if my memory serves me right.

Up until now I had never worked with sheep, but while on this farm a relative of my boss decided to retire from farming,  I was sent over for a fortnight, to assist in the preparations for the sale, this was quite a large hilly farm, that had practically no hanging gates between the fields, it was my job to get the sheep in each day, not an easy task there being no dog either, during that fortnight I walked until I could literally walk no more, despite all this I somehow took some sort of shine to the sheep.

Summer /54 it was decided to make quite a bit of silage as an experiment, a new green crop loader was bought to lift the crop on to trailers, to cart it all in, a stack of silage was built eventually, near the hay shed, the whoe crop was  pitched off the trailer by hand forks, this work was done in glorious weather, we had one five acre field left, when the weather forecast, such as it was then turned against us, a gamble was to be taken, I was told to mow this last field as quickly as I possibly could.

So all corners were rounded and everything was done to speed things up, when the field was cleared there were quite a few bits and pieces of the crop left standing here and there, especially on the corners and the field looked a bit rough, but the main bulk of the crop was safely harvested before the rain settled in.

Next day our elderly next door neighbour called, and the first thing he asked was, who mowed that last field? "Well", he said on being told, "It's not so bad, John, there is a full moon next week, you can take a scythe out after dark, to tidy it up young lad, at least nobody important will see you doing it!"

During September 1955 I left this farm after two years of very useful experience, and happily got married to a local girl.

Up till this time I had thoroughly enjoyed working living and learning, on these three farms, as said I had gained valuable experience as I went along, each farm was so different, yet they were all successful, each in their own traditional way.

# CHAPTER FOUR

Farm work on a larger scale
Fattening female Cattle only
Stopping Cows jumping over hedges,
and cooling them down.
Feeding Cattle better food,
Why I had one case of B.S.E. years later
Losing my Broad Beans.

Having rented a small local cottage to live in, I now found employment on a larger farm, and travelled some three miles to work each day, carrying my food with me.

This farmer was an elderly bachelor of 72 years, a cool and shrewd fellow; he regularly attended three local cattle markets each and every week, buying up all available barren and reject cows and heifers.

Living with his maiden sister, he was one of the most contented men I ever knew.

He would never buy a male animal, as he found they were slower and more expensive to fatten. This farmyard had a six-foot high wall all around it. The lorry people would deliver the animals and leave them in this yard; usually arriving with their loads during late afternoons, as many as 20 animals of mixed sizes at a time, all new arrivals totalling around sixty a week were intentionally left on this yard without any food till the following mid-day.

Being all rejects from farms, a large percentage were in calf, though they should not have been, many were horrendously wild and nervous, and positively dangerous to approach even.

Each week a few regularly jumped over the wall into the second lower yard, simply by doing this they had identified themselves as habitual hedgers.

Poor hedges and fences had allowed them to come and go, or roam, as they wish, a very bad habit that they had developed.

These animals had by now become such a nuisance to the previous owners, they were only too glad to get rid of them, simply by selling them on, and they now found themselves on our yard.

During my first week, my new employer told me, with a twinkle in his eye, that the secret of success was always to buy what farmers wanted to get rid of, and if you are selling always try to produce animals that everyone wants, and let them fight for them, and he certainly did this.

He was always buying what the dairy farmers did not want to keep around the place, and when he had fattened them, all the dealers vied with each other, to buy them for the meat producing abattoirs.

In my very first week on this farm I realized my previous experience, up to that point had been with rather domesticated, or quiet animals. I was now in for a sharp awakening, with a new large learning curve looming ahead of me, and believe me I had to learn quickly!

There were two of us employed on this farm, our Boss was very fair and straight with his men. I was to work, from 8am to 5pm, for my wage of £12 per week. The other employee worked from 6am till 3pm and we were both told to make sure we finished our work, and leave everything ship-shape on time, as he did not pay overtime, unless he expressly asked us to do so.

Routinely, my day began by taking the milk all in ten gallon churns, down to the roadside to catch the milk lorry, and afterwards it was my duty to wash and sterilize all the milking paraphernalia and so on.

My co-worker would already have done the morning milking before I arrived each morning, and would normally have gone home for his breakfast.

We then fed and inspected all the stock on the whole farm, before starting to sort out all yesterdays new arrivals. The ones that had gone over the wall into the lower yard, would be driven into the cowshed that had a row of twenty metal yolks.

Some food would be offered in the troughs, by now hungry, these cattle would unwittingly reach through these open yokes, which we then closed around their necks.

They were then able to lift their heads up and down, but there was no escaping, the wildest ones would be left to their own devices for a few hours.

Each one had to be stopped from jumping hedges, and without fail, not one ever jumped a hedge again!

Each animal had one regulation identification metal tag, near the tip of one ear, another tag would then be put in the opposite ear, a piece of wire could then be passed through both tags to pull their ears back together, over their necks, they could now not move their ears forward at all.

Quite often the wildest ones would lie on the floor in a temper when we did this. They would eventually be turned out into a large Orchard of half an acre, and would immediately go beeline for the hedges, but not one ever attempted to get out over a single hedge. They would simply stand nearby, looking foolish, until they eventually decided to move off slowly and graze.

The first time I witnessed this I could not believe what I had seen, after four or five days the wire could be safely removed, and not even one of these animals ever tried to jump a hedge again.

This experience definitely proved one thing, that a cow is not capable of making any decision with her ears tied backwards over her neck. After another day or so, their ears could be released, and they could all be safely turned out with the other cows, never to stray again.

The Orchard was hardly ever empty, as many of the new animals had this treatment each week throughout the year. All new arrivals were assessed carefully, as many would be fairly near to calving, these were kept together. Thinner ones would be separated and turned out to grass, the fatter ones would be put in the finishing shed.

The whole object of the farm was to produce finished beefy cows, a large number were kept outside all the year round. This was possible as the farm was a dry, sandy loam. Each week, animals would be taken to market, as well finished cow beef, while new animals also arrived regularly.

Throughout the winter we would feed a large number of cattle outside, but always in bunches of no more than twenty together in one field.

They would be fed on Silage but also some Hay, and dry feed, we were instructed to feed these animals in fairly large circles, and never in straight lines, this way the cows would cross over the centre of the circles, instead of walking in straight lines and treading or wasting the fodder offered to them.

Personally I have never seen even one instance of feeding being done this way, even when watching farming programmes on television, and in the latter years I don't miss many of them, whether they were feeding horses, cattle or sheep, but this simple method certainly saved a lot of feeding stuff being wasted, especially in wet weather when the ground was rather wet.

The cows that were milked were a motley lot of all sizes and colours, some would calve very young, and some would be extremely well grown being as much as four year old, like elephants, determined to kick your head off, when you tried to milk them. I always used a pincers in their nose to tie their heads up quite high, on to the vacuum pipeline, this would stiffen their spines so they could not kick too much with their rear legs.

Extreme cases would even be milked while lying on the floor. The pincer treatment in their nose never caused them to withhold their milk, a stubborn cow can easily refuse to let her milk down, but usually after a few days, they would accept the situation and cool down.

The ample good feed always in front of them made things much easier, but because they were not put back in calf they milked exceptionally well, and they were very well fattened, at the end of their lactation and would be sold off immediately as fat finished animals.

Each Friday, regular as clockwork, at eleven in the morning, I was to fetch from the local co-operative store, a standing order that included rolled barley, flaked maize, crushed wheat, rolled oats, dried sugar beet, palm kernel and molassine meal and a few bags of various minerals, in all a full six tonnes in weight, in fact the load was as much as the new diesel Fordson Major tractor of the time could handle, and on arriving home, I would reverse the loaded trailer into a large barn, just in time for lunch.

Friday was the boss's day at home, so after lunch he and I would tip these feeds into heaps all of correct proportions, and mix them thoroughly with shovels, then fill all this mixture into steel bins around the walls.

Every cow that milked, regardless of her yield, was given a bucketful of this well mixed high energy dry food twice daily each night and morning.

I never, before or afterwards, saw cows milking like these. Often during the winter months these cows would come in early morning, with six inches of snow on their backs, but due to the good feeding this didn't seem to hurt them the least little bit.

After a few weeks, I ventured to query the cost and sense of all the labour, mixing this feed each week. Surely, I ventured to suggest, it would be cheaper and easier to use compound cake, thus avoiding all this time consuming work. The boss smiled at me, and wiped his brow with his cap, and this was his answer.

"Young man, what you just said is absolutely correct, as far as cost and time are concerned, but I tell you this, and never you forget what I say, only a complete fool would think of feeding his cattle with anything, that he cannot be absolutely sure of what is in it." In that one long sentence I had been given the best advice I could ever have wished for, that proved to be correct time and time again.

Throughout my whole life thereafter, I have never even bought a single bag of reconstituted cattle cake. Except once many years later, I regret to say, I did buy two small bags of high protein pellets for my calves, I did this to try to boost their performance a little, and it proved a distressing disaster for me six years later.

I had sold a very promising pedigree young cow to a friend and neighbour of mine, but alas within two years this animal went down with B.S.E, he did not seem to be too concerned, as he got well compensated for his loss, but I felt sick for weeks afterwards, one mistake, just buying a little bit of extra protein, this really made me feel ashamed and disgusted with myself.

I was also extremely annoyed with the two Ministry men that visited my place in consequence of this; they even wanted to know exactly the spot where this animal had been born if you please, even though the animal was over six years old.

As I always calved my cows in a calving pen, I could show them exactly where this had happened, but I felt quite annoyed with them, as they could not explain why they wanted to know, and what this had to do with this disease after such a long period of time, and they also visited a second time with the same question!!

However, continuing with my story, one February evening after work, as the days were drawing out, I was given permission to cut some bean sticks, I was warned that there were some deep boggy areas where they were growing.

Seeing me take longer than he thought, necessary and darkness approaching, the boss came looking for me, quite concerned. "I was afraid of the bogs," he told me. I pointed out that there were cattle everywhere. "Oh yes," he said, "But cattle have an in-built sense of this danger, even in the dark, only once I had a cow stuck in the bog and that was my fault, because I rushed them too quickly."

In addition to all the cattle work, we also grew three acres of potatoes, and a few rows of broad beans, this particular year we had completed the potato planting, in rather dodgy and showery weather, having finished we then realised that we had rushed and completely forgotten to plant the broad beans, so there was nothing for it now, than to stick a bean seed in here and there along and amidst the potato rows.

Our boss was very kind to his workers, and we both had three rows of potatoes each for our own use, I decided to soak my allocation of bean seeds to speed up their germination, the boss and my fellow worker did not bother, so they planted their beans in dry, whilst mine were soaked and already swollen when planted, in due course their beans came up well and even, but poor me, I never saw a bean, this was a complete mystery to me at the time, and it took me a youngster a few years to understand and solve the problem.

What had happened was this, while their beans had absorbed moisture after planting, the natural swelling of the seed in the damp ground, had held the seed firm in the loose soil, whilst the roots pushed themselves down, my seed already swollen when planted were loose in the soft ground, so when the roots of mine pushed themselves down, they easily pushed the seed up to the surface, and the crows each morning had a whale of a time picking them up and playing with them, needless to say I never again soaked seeds prior to planting after that.

This was probably my first well learnt lesson in horticulture, (a posh word for gardening!) never to meddle with nature too much, doing what I did caused me some embarrassment at the time, but my loss was small compared to what I eventually gained, when I finally realised a few years later on what, and why, this had happened.

During the few mornings when there were no new cattle to sort out, my task would be to walk around the whole farm to inspect each and every animal out in the fields. I was to be extra careful with the newest arrivals, as the farm was prone to what we called 'Red water disease', caused by certain ticks or insects.

Affected cattle would start looking rather off colour. A day later, their urine would turn red as blood. I was offered half a crown 2s-6d, (12.5p) for every case I spotted, as early treatment was essential to save the animal, this incentive regularly amounted to an extra £2 a week, in my wage packet.

I decided that I would chase each suspect for around 100 yards or so, this had the effect on the animal to pass urine, which was either coloured or clear. Some days I would spot up to four with this condition, this early diagnosis meant that the animal recovered to good health so much quicker.

These ailing cattle would be brought in and dosed with black treacle while waiting for the Vet to come and treat them. This disease would never occur a second time on any animal and is nowadays more or less extinct. By going around all the animals, then going back with help to bring them in, we were regularly walking for most of the day.

During the early winter a few lambs were bought in to fatten, each time we had a snowfall we were instructed to fell a tree that was well covered with Ivy, our Boss was a big believer that a little Ivy did the sheep a power of good, the sheep certainly cleared the chopped up tree of all its Ivy very quickly, so there must be something in green ivy that the sheep recognised they wanted, of course the timber came useful for the house fire later on.

An animal would often be found to be missing; this caused very little concern for the first few days. Enquiries would eventually be made around our neighbours, but often the local lorry man knew where the animal was before we did, and he would regularly collect them back, at this time Horsey people only used small horse boxes, no farmers yet had their own cattle hauling trailers.

Early spring on the second year that I was on this farm, my Boss approached me and said, "I have ordered eighty tons of Basic Slag that is going to be delivered next week, and I am wondering whether you think that you can manage to spread it, as it is rather a lot of extra work,"

I replied that it was possible to do this, but on one condition that he buy me a full bag of builders sand to make things easier.

Each morning and afternoon I had to load about five tons of this, all in one hundred weight bags, on to the flat bottomed trailer, so to haul it out to whatever field we were sowing on the day, this was now very much easier as I was spreading a handful or two of this sand on to the trailer, this was enabling me to push the bags from the back of the trailer right on to the front ,without jumping up and down each time, the bags were actually rolling on the sand. After a few days the boss was rather quiet, and one lunch time he said to me that he was coming out to the field with me just to see how things were going. He was surprised when he saw me sowing round and round the field rather than back and fore, I explained that I thought it was quicker this way with less turning around, and he agreed.

Eventually the field was finished and back to the yard we went. "Well well," he said, "I must say, you have worked well this afternoon without wasting a minute, we have been out for three hours without a stop, but there is only one and a half hours registered on the new tractors hour clock?," I did my best to explain that the tractor was recording an hour in one hours time, when it was being driven at eighteen hundred revs constantly, and as our work only needed nine hundred revs, it would take two hours to register one hour. "Oh well," he said, "I don't understand it, but I know that you did your best all afternoon."

Apparently he had been checking the hour clock for some days and had convinced himself that I was slacking, next day the usual smile was back on his face, his sister told me the following day how he had been concerned, but that now he was happy, although he still did not understand how the tractor hour clock worked.

I stayed on this farm for another 18 months and then took a job on a more conventional dairy family farm.

This farm offered a wage of £12.50 a week, plus a cottage to live in free of rent, for as long as I worked there, this cottage had a large productive garden that I used to grow flowering Chrysanthemums in, during the autumn the garden looked a picture, with all the various colours, one evening our landlord, also now my boss, called round and when he saw the garden, he looked shocked and said, "Dear me, what a complete waste of good land." My wife had by now learned to drive our old 1931 Austin 10 car, so that she could deliver the flowers we grew to the county town florists, but alas towards the end of the season, the payments were slow coming in, and I soon realised that I still had a lot to learn once again, especially when I was away from farm livestock.

My new job was very fulfilling, but it came to an end when I resigned towards late August 1958. At this time I was as experienced as I could have been expected to be, for my age regards looking after and caring for farm livestock, but I was ever so conscious of the fact that farm finance was not my strong point, as up till that time no farmer that I worked for had ever discussed with me even the smallest detail of anything to do with the money side of things.

# Chapter 5

Now Confident to become Self Employed.
Dealing with the Supermarket.
Receiving good genuine business advice.
Realising the importance of Soil and Grasses
Why use Epsom Salts.

As I was now 24 years old, happily married with one child. Reasonably ambitious and confident, I had been restless for quite a while, and would have very much liked to start farming for myself.

My young wife and I had looked at smallholdings since 1956. These readily sold at around £50 an acre. Cows were priced around £40 to £50 and milk sold at 2s-0d (10p) a gallon or 2.5p per litre.

On paper it was impossible to justify borrowing a little, and start farming even on a small scale, simply because at the time we, my young wife and I, were completely ignorant of inflation as such. During the 40's (including the war years) prices had remained stagnant. Farms in 1939 had been sold for £33 per acre and cows were between £7 and £14 each.

Not being able to see our way ahead in farming, we therefore decided to buy a roadside cottage with three quarters of an acre of land with it, and as said this was done without a single thought as to the benefits of inflation, in fact this had never even entered our minds.

So on the 21st June 1958 the purchase of this small property was agreed, with completion of the purchase to be in late August. I was very lucky to get a private mortgage with a local elderly lady, that I had helped occasionally, by doing odd jobs around her smallholding, as well as helping her with the annual hay harvest.

Banks did not give mortgages to ordinary working people at that time, but by late August we were lucky to able to move in and live in our own little home.

We gradually cleared the ground, but it took me a full three weeks, just to clear away all the empty bottles I found in the undergrowth of immediate garden area behind the cottage, as this had been a roadside pub for centuries before. It was also just across the road to the site where cattle had been collected and shoe'd prior to being taken by drovers to far away markets in olden times.

Little did I think that this was a natural trading place, but this soon became obvious, a sort of luck a chap gets once in a lifetime. We then slowly developed this roadside plot into a plant nursery and flower shop.

At first we grew and sold mostly vegetable plants, these sold on a very small scale, we had a few pansy flowering plants that I simply could not sell to the local people, I decided to get rid of these by giving one plant to each customer that spent a few shillings (5p) with us for vegetable plants.

One lady looked at the plant I gave her, and asked me what she was supposed to do with it.

I well remember telling her to plant it somewhere where she could see it, and if she kept the sun shining on her back when she looked at it, I could guarantee that the Pansy would always be looking at her at the same time, and this was quite true, as all flowering plants always look towards the sun.

Early in the year 1959 I grew quite a lot of Sweet Pea plants, I had put the first batch outside to acclimatise or harden off, before selling them, these were close by the kennel that housed our family pet, a young Pembrokeshire Corgi bitch, coming into season she had attracted the attention of nearly every dog in the village, each one had decided that the sweet pea boxes were warmer to sit on, and as I arrived on the scene one early morning, there was hardly a trace of any plants left to see.

Just my luck, I thought, this was a total disaster, but a few days later they, the plants started sending out side shoots, and eventually proved to be a much better crop, it is now common practice to nip out the growing tips after two leaves of the young Sweet Pea plants, as they bloom much better on the resulting side shoots, that was truly a little bit of beginners luck that we all need, I am sure.

Spring 1959 we decided to plant some potatoes on the top half of our bit of land, as this would help to improve the soil.

When it came that we were ready to open the drills, my father-in-laws Ferguson tractor was busy, and as there was a cloud on the horizon, I hitched the double-tom, (a drill opening plough) on to the bumper of our Austin Ten car, with my wife at the wheel we soon opened the drills, much to the merriment of the public, that were driving past on the busy A40 road that was our south boundary.

For the first two years, I also worked three days a week on numerous local farms doing some relief milking, cutting hedges and so on. At the same time, and whilst I did this work I learned a lot regarding the local farmers' beliefs, traditions and foibles.

For example, one farmer I worked for on odd days was an academic, he had opted out of his previous life style to take up farming on a small scale. He was always extremely thorough, and his outputs were tremendous. Each time I arrived there would be a list of jobs for me to get on with. Firstly I was to go out and spread all the fresh dung pats, this he did regularly every day without fail.

When making hay he would cut any old rough grass he could find, he even cut a few miles on both sides of the road, and he would mix all this in with his best hay, explaining to me, that this extra roughage was as important, as the better value grasses, for his milking cows.

In the mid-fifties he shocked all his neighbours, when he started sowing fertilizers at regular intervals throughout the summer, he became the first one to do this in our locality.

On one occasion he was selling a few pigs in our local mart. The trade was slow. The Auctioneer stopped struggling and said to him, "I think this is the best I can get for you today, I feel you should accept," in a flash he replied "Sorry, I think I should have more than twenty pence a pound for them."

That Auctioneer was not happy at all with clients that were able to count like that, it was the reaction of that Auctioneer, that made me realise perhaps for the first time in my young life, how people that should know better, consider themselves to be superior than most of their fellow men.

By today the time of writing, this is what I believe and say with conviction. "Never trust completely, anybody that is better dressed than yourself," hard experience of dealing with so called professional people, have prompted me to think that way, an ordinary sort of rogue is a rogue, an educated rogue is something rather worse, believe me, or wait till you get bitten. To be fair this may not be completely correct, but I am also sure of one thing, it is not completely wrong.

A year or two on as our nursery was slowly establishing itself we would have more flower plants than we could retail ourselves, so I would supply one shop that was part of a chain, like our supermarkets today, at first I delivered their weekly order in full, and they would regularly haggle over the price, soon I was only giving them 75% of their order, and surprise surprise, their concern then was when could they have the rest of the order, with never a mention regards the price.

Spring 1960 we decided to attend a town market, on Wednesday and Saturdays, but we soon decided against this option, as I could not be in two places at once. My wife and I worked hard and long for the next few years, as our little nursery progressed steadily, ours was the only plant nursery at the time in our area, the nearest one at least 20 miles away.

Our market town was 13 miles away and I would often seize the opportunity to do any nursery deliveries in that area on a Thursday, which was also the store cattle sale day, this allowed me a quick visit, which I restricted to around half an hour or so, just to satisfy myself, as I still had considerable interest in cattle.

During the first year at the nursery I was privileged to have some excellent business advice from my elderly customers and friends, this first spring I planted a few Dahlia flower plants, just inside my boundary hedge on the side of the main A40 road, into Pembrokeshire.

I decided I could do with a few more for next year, so I visited the Three Counties Show, at

Carmarthen, where a prominent nurseryman was exhibiting a good selection of plants, and I ordered a few for delivery next autumn. He was curious as to why I needed so many, and I told him we were gradually starting a small nursery.

He was very helpful, and could not do enough for me, he told me, that he had spotted my few plants when he drove past that morning, and that we were starting at a wonderful location, and he wished us very well with our venture, he finished up by saying that we would succeed, as long as I kept my feet on the ground, and however successful I became, to remember my beginnings, and whatever I did in future, not to let myself become even the least bit big-headed.

A retired country shopkeeper, also wishing us well, went on to tell me the following. "Your priority should be to make sure you have One Penny more in your account, for each and every day, that is seven pence for each week that goes by. If you can achieve this," he said, "you will be on the correct track and you will not fail."

Being extremely busy it actually took me a few months to realise what had been said, but what he meant was, don't over commit yourself financially, never buy anything on HP, and make sure you always pay your bills on time, with a penny more left each day, so completely avoiding debt.

This proved a very hard act to follow; in fact often practically impossible, but this was probably the best advice I could have wished for as a young man starting a small business, and hopefully, I have never subconsciously forgotten those pearls of solid wisdom, now this today may seem old fashioned, but I think it is wise for a youngster starting out in business, to have a few years experience of dealing with people before he starts borrowing too much!

About three years on, my part time farm work ceased altogether. Each year till then I had erected another green-house, so thereby expanding carefully. From 1961-1981 the nursery occupied all my working time, but I often found myself mulling over, and perhaps dreaming of the highlights of my previous years in farming.

Now as a rather raw nurseryman, growing a multitude of plants, I quickly realized the importance of correct growing conditions, above and below the ground surface, each family of plants demanded individual, and often very exacting soil conditions, and I gradually realized the reason for some of the individual odd behaviours I had seen in cattle, over the previous years

Grass is, and probably always will be, the most important, easiest and the most economical plant to grow, it sustains our farming industry, that produces a large percentage of our food, firstly grass satisfies the appetites of our animals, grass also absorbs all the dissolved nutrients, trace elements etc., that are available in rainwater, as well as in the soil within the area that its root system can reach, grass also absorbs nutrients quicker through its leaves when it actually rains, rainwater is simply sea water that has evaporated around our coastline, and this contains weak solutions of the important and essential Trace Elements, that are important to all men and beasts.

I say this with conviction, after seeing how grass grows under a covering of snow, when it has been hanging around for a while, I have also heard farmers refer to snow as the poor man's fertilizer, yes grass collects the afore said nutrients and conveniently delivers them all, to our animals Gut, so they can naturally make use of them one and all, for their benefit, and therefore indirectly to benefit all humanity.

One example of grass and plants absorbing nutrients through their leaves is the quick greening of grass after it is sprayed with a weak solution of water and Magnesium, known as Epsom Salts, simply make a mixture of one level teaspoonful dissolved in a glass of warm water, add this to two gallons of cold water and spray an area of grassland, or just use a watering can fitted with a rose and slosh it about, this will almost immediately green up the treated grass area, and definitely it will be noticeably greener within a very few days, this helps the Chlorophyll in the grass to green up.

Spray your whole farm occasionally, for a minimum cost, and you can say goodbye to Staggers and Milk fever, for quite a few years at least. By greening the lawn this way without using fertilizer, you now need to mow less frequently, all grasses absorbs magnesium  quickly through their leaves, but very slowly through their roots.

It was my nursery work that made me realise the value of Magnesium, many varieties of plants would look a bit yellow, especially after a spell of cold weather, but the faithful Magnesium never let me down, it certainly made all plants look much greener and better, and therefore more robust.

# Chapter 6

Does the use of excess Lime cause Cows to be nervous?
Why value Worms, and the ideal Worm count.
What do the letters P.H. mean.
Leading all Cattle easily.
Avoiding Laminitis in fed Cattle.

The previously mentioned cows, on the first farm I worked on, (Chapter 1) were spooked by the sudden appearance of two goats, and stampeded in panic, all bar the Red Devon cow that kept calm. Now this cow had only recently joined the herd, and she always grazed the hedgerows by herself, in fact, she was avoiding the grassland, as much as possible, having I guess recently come off better land, so therefore was now probably suffering while adjusting to the poorer soil conditions, compared to what she had been used to, we are now again talking of farming as it was done in 1949, when farmers, and most especially this particular one, were only beginning to use Artificial fertilizers.

Most farmers at that time were only using fertilizer for one sowing annually, to kick start and get a better crop of hay, in June or July, as the weather allowed.

My first boss, having married a little bit above his station, was always keen to impress his in-laws with his farming ability.

Small scale as it was, he was using a lot of Lime, it certainly greened up the pastures and made the herbage sweeter, and more palatable for the animals.

Also using what was then known as "Burnt Lime" or "Hot Lime," that was delivered in lorry loads and tipped up, we would load this bulky material onto the cart, and place four shovels of it, in heaps, and straight rows across the field. Each heap and row was placed seven yards apart from each other.

This was then left for 3 weeks or so, to slake, by allowing it to absorb the rain that caused it to swell into a large heap, it was only then spread evenly, by shovel, over an area of 49 square yards off each heap.

This lime was much stronger than the carbonate lime used today, therefore, the ground was grossly over-dosed, in comparison to today, Basic slag was also popular, and this probably contained a high percentage of Iron, on reflection, the cows were nervous wrecks, because the natural Trace Elements, were locked up, or neutralized, some way or another, by this excessive use of lime and Iron.

Each year this farmer would plough a few acres to grow some oats and green crop. I know that he believed that this ploughing was bringing the lime back nearer to the top of the land, and quite rightly so, because after lime gets washed down deeper than the grass roots, it is lost for ever as far as the farmer is concerned.

This may have been a correct and useful procedure, but lime is also brought naturally back to the surface, in a continual process in fertile soils, by the earth worms that are continually bringing Lime back to the surface.

But for now, let's first think about the cows in question, the white cow, that was seriously ill, she certainly recognised the difference in the soil, that was offered to her from the middle of the hedge, she quite naturally detected by smell that the cure for her illness was in that soil, and promptly ate a

64

panful regularly twice a day for a fortnight, this alone gradually fully restored her good health.

We mere humans may also have that ability as well, if we only could realise it, for example why do we sometimes develop a sudden craving, or a need to eat or drink in excess, are we subconsciously looking to correct deficiencies in our bodies? Perhaps medical researchers should look deeply into this.

The Red Devon cow also tried to avoid these deficiencies as said. The Friesians were definitely more vulnerable as well, as they had not been in this country but for a very short time, their native soils in the Netherlands, on the northern European sea shores, are probably much better as regards natural nutrients and Trace Elements, simply due to their location, practically below the level of the North Sea, after all, marsh lamb is known to be better tasting. So now, let's see if we can satisfy ourselves that maybe some things I have touched on may be true.

I am sure most farmers have seen cows eating bits of old newspapers that may be around. Why is this so? One of the most important Trace Elements that is natural in soil and should always be available in grass is Cobalt. The largest customers for mined Cobalt is the printing industry, all newspapers are lightly dusted over with Cobalt to stop the pages sticking together. Our human bodies absorb some of this through our skin; this is what makes our hands black or grey when we handle or read them.

Even young calves can recognise the presence of this vital element and enjoy eating newspaper as well. To check this out all the stock farmer has to do is to tear a daily newspaper up in strips of an inch or so, and tie these individually, on to gates and fences or where ever animals can have access to them, I have yet to see animals ignoring and not eating paper strips.

Small amounts of newspaper will not do farm stock any harm; any farmer that can say his cows or sheep will not touch newspaper can congratulate himself with complete justification, for having a near perfect trace element regime as far as Cobalt is concerned, if grass land is short of Cobalt it has less ability to retain Nitrogen, due to a lack of the Vitamin B 12.

Cows and calves also love to chew on any plastic that is lying around, including plastic string, this is because Selenium, another very important Trace Element is included in the making of most plastics, and our animals are clever enough to know of its presence and take advantage of it.

Today, and by now gardening in retirement, I bury newspaper deep down in my Runner bean row, as beans also benefit greatly from extra Cobalt and extra Vitamin B 12 as well.

Let's get back to the lime, and how does this recirculate to the surface?, after being washed down too deep, mostly by the rain. The farmer's best friend, the earthworm, does this for him for free. Why is the worm also dependent on lime?

The worm is a very active creature, but he must have damp conditions, the drier the weather (and therefore the soil) the deeper he goes. Looking closely at a worm, he looks very delicate and tender, he burrows himself down during the day, to avoid the warmth and/or the light of the sun, as he cannot tolerate much of this.

In darkness he comes up to feed, mostly on dead vegetable matter and any old rubbish, such as old dead grass, which having eaten he re-cycles this as worm casts, thus continually improving the soil.

Now lime or limestone soil is always rather soapy and slippery, this lime gets mixed up in the soil and water, the worms also eats some of this mixture, and separates some of the lime and stores it in the visual sack that is prominent somewhere around their middle, then as they travel up and down through our (and their) soil, they cleverly exude (or sweats) this lime and by so doing they reline the insides of their tunnels, simply to keep them smooth for their own comfort, as they continually travel up and down in the soil, by continually doing this they are actually re-circulating the lime that has naturally gone too deep, back up towards the surface, so that it becomes available again nearer the surface, this action alone is a great benefit for the farmer.

The hollow worm tunnels also allows rainwater into the depths of the soil, and as this water contains soluble oxygen, and to a lesser degree some Hydrogen, (the fact that fish live in water is proof of this), it is this air that is vital to the grass roots, as they have to breathe to grow at all, it is a fact that without worms and their activities, the farmer's land would be mostly dead and infertile.

I mentioned Hydrogen in rainwater, when we test for Lime values in our soil, we use a PH scale from 1 to 12, with 5.5 to 6 as ideal for growing grass, the letters P H means Power of Hydrogen, or Potential of Hydrogen, this is determined by measuring the mixture of this Hydrogen and available lime in the soil sample.

Farmers certainly ignore their worms at their peril, just like chickens, each worm has a gizzard, this is an organ in which the worms store rough grit, up to 1/20th of an inch, that is used to grind down their food just like chickens do. When the topsoil is saturated with rainwater worms spend a lot of time on the surface on dull days, but they are mostly active as said during darkness.

Worms also benefit our soils, each time they re-enter their tunnels, because as they go down, so they compress the air downwards towards the deeper roots of grasses, and this air is absolutely essential for the roots.

The ideal worm count for fertile soil is at least thirty per square meter, that is equal or near to one and a half million (1,452,000) worms per square acre, or 3,600,000 per Hectare. Anything less and production will be more expensive than it should be and therefore, less profitable. Before farmers became aware of the value of using lime it was only the activity of the worms that kept the soil fertility such as it was.

A quick way to count your worms is to soak a square foot of ground with a fairly strong mix of washing up liquid and water. The worms will immediately come up for air, this should preferably be done on a wet day, so that the sun will not damage them.

In decent soil, you will easily find the numbers I quoted. If you do, you are a fairly good custodian of your soil, and your own well-being, any less there is room for improvement, which would greatly benefit the farmer's health, and that of his stock, and equally important his profits.

However, worms do not surface during frosty weather, to prepare for this, they take down stocks of vegetable matter, such as dead grass and store it quite deeply, they also take down whole leaves fallen from trees, to sustain them during such cold weather spells.

I have personally during late autumn often seen an oak leaf standing up half buried on a patch of clean soil, each day it was getting a little deeper as it was gradually being pulled down by a worm, or so I presumed.

Speaking of cold weather, I have also found that the best time to reduce the vast populations of slugs, is just after a thaw has set in, slugs also go deep to avoid frost, but when they re-emerge, they are usually starving and very thin indeed, they will eat any green herbage and whatever else they can find, so this is a good time to bait them.

Let's take a few moments to remember a few happenings in the life of this Farmer's Boy. Now years later by simply thinking about the young calves and how they remembered the barb wire incident on the first farm I worked on (Chapter 1)

I reasoned perhaps I could turn this to my advantage.

At first I simply tied a string loosely, around my calves necks, fairly soon after they were born, then holding on to them occasionally as I passed by, I found that they soon stopped struggling.

I then tied them to the top of a gate, even during their first day of life, and they would soon be lying down again, not the least bit bothered. Soon I progressed to putting a little homemade collar on their heads, these were just long enough so they could step on the loose end with their front feet, now they would struggle a bit until they moved that front foot, but quite soon they were all leading without hardly any extra effort on my behalf.

Then, at eight months, or whenever I needed to handle them, there was never any more trouble, and certainly no more struggling and pulling, and importantly much less risk of injury for myself and the animals. One wet afternoon I went a little further, and tied two very young calves together leaving them to pull and tug at each other, they too soon gave up and both laid down contentedly, all this was very little effort, and it resulted in a much easier life for me later on.

The next year I had eight calves in a shed, and I noticed a calf suckling at one of their collars. Thinking as a nurseryman, the answer was simple. I mixed some 'Alum' powder in water and dipped the collars in this, and I had no more trouble.

Alum powder has a very bitter taste, on the Nursery we used to spray our plants with this, to stop the birds pecking at the flower buds, and the calves found it bitter likewise. This powder can still be ordered in any good chemist today. Probably lemon or any bitter substance would do the same.

The question now is this, is a little time and effort early on worthwhile, it is certainly, if you show cattle and sheep, and definitely if you value your time and effort spent training and struggling with larger animals later on.

Today a few years later I understand that leading very young calves is now standard practice in the Netherlands. Perhaps the farmer's boy was not so daft after all!

Another important benefit of leading calves as young as possible, is that it avoids bruising their feet when they struggle, as they do when they are heavier and larger, this damage does not become apparent for some time, and invariably develops into serious foot trouble some time later on.

This is often so with bought in young bulls, of course bulls can also suffer with bad feet due to gross over feeding, a condition known as Laminitis, this is caused mostly by feeding too high a percentage of protein, this seems to have the same effect on elderly humans that suffer from gout.

Again by leading young bulls early you have greatly reduced the stress that they suffer each time you decide to handle them, when one does not seem well, and even more so when taking them to be sold.

Everything said here about leading young cattle is also applicable to sheep and horses, as they never forget either, I have often seen prominent sheep exhibitors at local shows, picking up a well grown lamb and carrying it in to the ring out of frustration, when it fails to lead quietly.

Having just mentioned Laminitis, this can be easily avoided, by simply feeding your animals with well dampened or wet food, even wetter the better, over the last fifteen years of my working life I can assure you, that I never fed any animal, intensively or otherwise, with dry food, be they chickens, sheep or cattle, and I shudder to think how much I lost out on over my lifetime by doing so previously, of course it is a bother, but it is also extremely profitable, in cold weather this dampening is done using warm water, especially if you are looking for maximum weight gains, as you do, when feeding young promising bulls.

# Chapter 7

Soil Fertility.
The use of Sand versus Straw.
Stirred Slurry, good or bad?

Let's get back to our soil. I recently became aware of a farmer who was becoming concerned, as a considerable acreage of his grassland had deteriorated so badly, that there were only a very few grasses growing per square meter.

The cause of the problem turned out to be that for a number of years, he had found it really economical to use sand as bedding for his cattle, instead of straw. The cheap sand in question proved to be what had been blown in off the seashore, and on to a car park during successive winter storms.

This material is sold off quite cheaply, but as it is windblown, it is a very fine grained sand, likewise is the cheap sand sold off by quarries, this stuff is so fine, it is of no use to builders, or anyone else, but it kept this particular farmer happy and his cattle are comfortable in the sheds, and everything was hunky-dory all around, the emphasis being on the word was!

But by now this sand has found its way out onto his fields, the essential air spaces in his soil were sooner or later completely blocked up, so that vital air could not penetrate, deep enough, to be of benefit to the deeper roots of his grass.

Had this farmer used a more expensive rough and gritty sand, some of the rougher particles would have stuck to the topsoil and possibly helped to open his soil up a little.

There is an old saying; - Clay on sand, improves the land, but sand on clay, will never pay.

The three main essential elements for good grass growth are Warmth, Moisture and Oxygen or Air, this farmer should by now be beginning to realise that things are going awry, and is probably still wondering what is going wrong, with his grassland.

The obvious quick remedy of course would be to plough the land up and reseed, a costly business in itself, let alone dealing with the soft nature of the field afterwards.

This will be only a very temporary solution, because that fine sand will still be there, and will very soon be washed down solid again, and I fear, will cause the same problem over and over.

Surely that is enough said, I look forward to meeting the gentleman who finds an economical way to separate this evil substance, out of the farmers soil, and get rid of it, without disturbing the Farmers production schedule.

I personally think of this sand as an evil substance, as that is what it is, when it has been mixed in with soil used for growing grass crops, sand is so dense when wet, the only natural place for it is on the seashore.

It is useful there to slow down the strong waves during heavy storms, and as far as farming is concerned, it is there it should be left.

I feel sure that the next ten years or so, this use for sand will prove very costly indeed, already it is costing a little, as contractors are already charging extra for pumping out slurry, if sand is used as bedding, quite rightly so as it does not take long for the sand, to wear out the innards of the large pumps they use.

Another argument, could be that there are large tracts of sandy lands all over the British Isles. Fair enough, examine that soil and you will soon realise the difference.

After many years of tree planting, over a wide area of West and Mid Wales, there is no question in my mind but that the best soil for any location, is the natural soil that is already there.

I make no attempt to apologise, for my remarks, emphasising the danger of using sand in the last paragraph, the better alternative has to be straw, and by today it must be possible to chop it up, so that it is suitable to use with the modern scrapers and pumps and other equipment thought and believed to be necessary today.

Whatever the extra cost of straw, this is soon cancelled out by the benefits it brings with it, the fertility taken off the land on which it was grown, for example, forty small bales of straw bought in, brings the buyer, the equivalent of one hundredweight of lime, or Calcium, that is already in available form for his animals.

So long as straw falls inside your gate you own it, give it to your animals, either to eat, or as ample bedding, they are only borrowing it, and they are now processing it for your benefit, and you the farmer have not lost it. By buying straw you are now investing your money into your soil, as opposed to spending it on infertile and innocuous waste sand.

Careful handling, and usage, of good strawy manure will improve your land immeasurably. Your faithful worms will rejoice, they will multiply greatly, and each one will work hard and faithfully for you, all to the farmers benefit.

Cost into the equation the resulting need for less artificial fertilizers, your eventual profits are bound to improve.

I mentioned careful handling of dry semi-solid manure in the last paragraph. This is very important. Misuse, by rushing and rough handling of large machinery, can also kill a large percentage of the live bacteria and microbes that are always present in all fairly dry and warm manure heaps, again there is no room or need, for anything dead on any farm. Live bacteria works for the farmer constantly. We all know that anything that stinks is probably dead, and is therefore of no immediate use to the farmer.

Likewise, well stirred slurry stinks because the bacteria in it has already died, probably drowned, in stale and dirty chemically contaminated water, or in simple words, a concentrated liquid once again deprived of vital air.

Having spread animal manure in this condition on to your fields, there will be a considerable time lapse, before the live bacteria that survive can replace the numbers already drowned, and also get to work on it, so to turn it back into a sweet and useful available form, that will benefit them (the micro-organisms) and of course the ever important worms, and all other soil life.

In my experience, the warmer the soil, the sooner the dung disappears. I found the month of August best of all for spreading dung on grassland

If I had fields free of stock at the time, a really heavy covering of dung could well disappear in under a month; the richer the soil the sooner the spread dung will disappear, simply because of the higher numbers of active Worms, and microbial bacteria etc.

Having said what I have about Slurry, I repeat, "If it stinks it is long dead", but this does not worry farmers today, their problem is to get rid of it as cheaply as possible, some even mix bag fertilizers into liquid slurry, and do the two jobs in one, by using a trailing pipe system to spread it over the fields.

If Slurry is dead, and therefore stinking when applied to grassland, with no air or dissolved oxygen in it, the worms come up immediately gasping for fresh air to breathe, they are then exposed to daylight and the sun, and are soon killed by exposure, then again where ever there is a tractor going continually, there is soon a few canny seagulls looking for a free meal, and they soon devour all the exposed worms, whether they are already dead or alive. I have stopped and checked large fields and have found an average of four fat seagulls that have stuffed themselves silly, on every square yard, over the whole area.

Compare this to driving out and spreading reasonably dry dung, this material stays on the top of the field, and the occasional shower washes it into the soil, a little at a time over a period, this the worms and various bacteria can happily cope with. I know keen farmers on exposed windy sites, who spread dry dung like this each spring so to give their grass a certain amount of shelter from the wind, so it can grow a little earlier in the spring, especially, if they are on light sandy soil.

As said the cheapest way of disposing of effluent will cut costs, and help short term profits, but if you wish, to leave a worthwhile Legacy to your family, look after your land and soil better at the present time, and certainly, try to stop feeding the Seagulls, and killing your worms.  Having mentioned seagulls, I can easily remember ploughing with horses, there was then no noise at all, it was never long before they appeared in large numbers to help themselves to worms turned up by the plough, it had to be their keen sense of smell detecting the fresh earth being turned up, or they had scouts out looking, and a wonderful way of communicating with each other, that we humans know nothing about!. With the plough they had to be quick to catch the worms as they were alive and soon disappeared into the soil again. But not so when they come up for air avoiding thick slurry, and it may be possible that seagulls are clever enough to identify bad smells, with what they think of as a rattling good free meal.

Farm land and soil I have found to be honest, the better you treat it naturally, the better it will reward you the farmer. Personally I think it is much better farming when the liquids are pumped out without stirring, this dirty water the worms and soil bacteria can cope with, I know from experience this does the land a power of good as such, I have only mentioned earthworms here, imagine how the millions of beneficial bacteria and other microscopic life in our soils are upset and killed when flooded with toxic airless liquids such as stirred slurry.

During the last few years I have kept a small wormery, this is simply a household bin fitted with a false floor and a drain tap at the bottom, fifty tiger worms from the fishing shop, and a small shovel full of soil were initially introduced, and then all our house hold food waste has been added each week, everything except orange and onion peelings, soon the worms started breeding and the next year the effluent was  drawn off quite often and used after being diluted, one part is added to ten parts of water, and used as a liquid fertilizer, with amazing results, its potency is quite remarkable, giving exceptional colouring and growth across the board to all the vegetables and flowers in my garden and greenhouse. The whole bin is now teeming with worms, and is very worthwhile keeping, the liquid produced is completely natural and organic I would say, one day the bin is nearly full of soil and the next it is half empty again, such is the activity of these worms. Recently a long standing friend of mine and I were discussing our wormeries, when he said to me, "Why were we so stupid as not to think of this many years ago, simplicity beat us my friend, and we both thought that we were quite capable, not to put too fine a point on it!." Considering what these few worms are doing for our small gardens, can we begin to imagine how much goodness one and a half million worms do per acre of grassland for the farmer, and are doing continually every day.

I understand that calculations show that annually several tons of new soil is produced per acre by the plethora of various soil life organisms, this is left on the surface of each acre, this does not surprise me as it does not take long for a chain harrow to disappear if left out on grassland. One final word regards the use of sand , I hardly think that any farmer/s I know today  would swap his/their fertile farm for double the acreage of natural and useless Sand dunes, if that is so, why carry the Sand dunes onto your soil? This previous chapter has certainly not been included to upset any individual, where sand is used for convenience or gain, eventually it will prove to be extremely costly, and a long term very costly mistake.

# Chapter 8

The Vicars House Cow and
My dance with Death.
One way to get Suckling cows to breed regularly.

Looking back once more, to the mid-1950s, the vicar in our local parish decided to keep a house cow, and soon Buttercup, a jersey cow arrived on the scene.

Our vicar had an acre or so of land with the vicarage, there were also two graveyards, a few miles apart, and some grazing around the local vestry. This cow soon became a local legend, often seen trundling along behind the vicar's car, from one location to another, and occasionally his two young daughters would be seen leading her along the quiet lanes quite happily.

Being well looked after, this cow provided ample milk for the family, although she would sometimes make her mind up, to go for a walk by herself, she was apparently quite partial to whatever was hanging on the local washing lines. Eventually, after quite a few years, the decision was made to sell the cow.

A neighbour soon bought her, and took her home to his smallholding nearby, and kept her along with the two cows that he already had. As he was working full time, he did not have much time to attend to his cows, let alone to pamper them.

One evening some six weeks later as I went past, I noticed an animal on the ground, covered with a tarpaulin. On enquiring, I learned that Buttercup had sadly and suddenly died for no apparent reason at all.

I myself felt quite sad, and I was soon convinced that she had grieved herself to death missing the family life that she had known. Years later I had good reason to remember this incident, as we shall see later on. During this period I remember seeing the largest lorry I had ever seen passing by, on its side were painted in large letters the word OMO, we all wondered what the hell did OMO mean. A few weeks later it was available in local shops as the best washing powder available, but it did not seem to be around for long.

In 1964 my grandmother sadly passed away, leaving my spinster aunt by herself on the local estate owned smallholding. As joint tenants, the estate management had kindly allowed them both to remain as tenants for their lifetime, if they so wished, this was in appreciation of my grandfather's fifty years service to them, as head gardener he had maintained the gardens, which were very important for their image, as well as looking after their woodlands that supplied the estate farms with the timber they required to maintain their buildings.

So by 1966, my aunt finally decided to sell her few milking cows, and only keep her two in-calf black and white Friesian heifers that were to be sold next autumn, with their expected calves at heel.

So I was given the opportunity to keep a suckling cow or two there for the summer months if I so wished.

At the next loal market, I was there looking for a suitable cow to rear a few calves. This was now my first ever visit to a dairy cow sale, and mildly surprised, I soon realised that quite a few of the cows on offer were stale, or not as freshly calved as I personally thought they should be.

The old saying 'buyer beware' sadly came to mind, even so, I soon fixed my mind on a smallish grey cow, that seemed suitable for my purpose. She was rather lean but looked quite healthy. When her turn came I managed to buy her and her calf for seventy pounds. I then bought a good beefy calf for ten pounds, so home she came with two calves.

I offered her some well dampened food, (as said this is much better than dry food) and it was obvious that she was not used to being made a fuss of. Her priority was looking after her calves. Normally cows reject a new calf that is introduced, but not this one.

By the next day it was quite clear that this cow was capable of producing more milk, according to what she was fed, and by the end of the week she was suckling four decent beef calves. I had even sold her natural calf and swapped it for a better one.

After a few days I decided it was safe to turn her out with her brood, but alas as soon as she was outside nobody was safe anywhere near her, so back in she came again, and we gave her the wire treatment, (i.e. her ears were tied back over her neck, as I mentioned previously) and after a few days out she went, calves and all, with no further problem.

She was now given our home mixed feed the other side of a gate, (this was for my aunt's safety) night and morning. She now looked like a dream come true, looking after and feeding, four healthy calves with ample milk for one and all. Once a week or so I would visit and each time I entered the field she was in, she would make a low mooing sound, and the calves would then follow her, and off to the furthest corner possible she would go, turn around to face me with her adopted calves all obediently behind her. I had never seen this done before, or after, or ever heard of it being done.

The cow and her family, were now grazing with my aunt's two in-calf heifers, and all seemed fine.

One Sunday morning in August, around ten-thirty, I arrived to find my Aunt not at home. I proceeded up to the field to find the cow as usual leading her calves away from me.

As I needed to have a close look at the cow's udder, I naturally followed her, but in so doing I went past the two heifers without really taking much notice of them. I was walking along, casually concentrating on the cow, when suddenly... bang! I was on the ground face down!

One of the heifers had charged at me from behind, and I had not even heard her approaching.

The first thing that I was aware of was that she was repeatedly charging at me, again and again, completely demented and bawling loudly, with her tongue hanging out! I suddenly realised that I was in a rather serious situation.

By now on my feet, she was so quick I could not dodge her, her horns were too short for to get a hold on to them, so desperate, I grabbed at the bend in her tongue with both hands and amazingly, she actually pulled me along the ground for at least five yards before I lost my grip. In the heat of the moment I grabbed her front foot and all but managed to throw her over, but all to no avail.

This battle went on for nearly half an hour, now aware that I was losing ground and getting nearer the bottom hedge all the time (which would have been fatal) then all of a sudden, I remembered the bull years ago.

On my second attempt I managed to sink my thumb completely into her eye, and Holy be, this stopped her! There she was now holding her head up high, turning round and round, obviously in pain, that by then I thought she well deserved!

I soon legged it out of the field and sat down for a few minutes in the hedge getting my breath back, aware that my thumb was paining me, but I was ever so thankful that I had rescued myself by using it.

After a while I noticed that my glasses and hat were missing, so I walked slowly down to the barn for a pitchfork, and eventually found them where the fracas had started. I half expected my lady to come for me again, but she was still smarting because of her eye. In any case, I was ready with the pitchfork and thankfully she thought better of it.

Eventually I got to my car, and becoming more and more aware that I had been in a right royal battle. Luckily my doctor was in his surgery when I called in, on my way home. He had seen me approaching through the window and met me in the doorway.

"Come in young man," he said, "Where the hell have you been, and however, did you get from there?" It was then that I realised that my clothes were in tatters, and all mucky.

The doctor diagnosed three broken ribs on one side and four on the other side, which was well more than sufficient. My whole body was fast becoming bruised, including my arms and legs, and my precious thumb was completely black.

In bed for several days, I wondered why I had been so slow, remembering about the Bull, and his tender eye a few years ago.

For days I mulled over and over in my brain, why this heifer had acted like this; a perfectly quiet animal up until then. Gradually it dawned on me, if it was possible that being in-calf her mothering instincts were kicking in?

The only answer I could think of was that she was already protecting and/or mothering the calves. Some weeks later, still with my pitchfork in hand, I ventured into the same field as her, and sure enough there she was licking one of the calves. So that was probably the reason.

The heifers eye was still swollen a little, and she was still quite apprehensive, circling around me and my fork at the ready, but she thought better of trying it again. A hard lesson learnt, in a very hard way, ever since I have never kept first time pregnant heifers and suckling cows and calves together, this heifer was one of the most dangerous animals I ever encountered in my life.

To this day I consider myself ever so fortunate, that I did not suffer serious health problems, after this very severe and deep bruising.

That animal was later sent for slaughter, her whole demeanour implied that she could never be trusted ever again. I seriously thought to give her a long period of the wire treatment, but it was definitely not worth the risk.

Reflecting back at that attack, I feel sure that had it happened to me ten years earlier, or even ten years later in my life, I very much doubt if I would have survived such a determined and vicious assault.

Now back to the grey cow. She diligently nursed her four calves for a full twelve weeks, when I decided to wean them off.

I then tried another two good calves on her, she took a shine to them immediately, bless her, and she reared these for another ten weeks, we kept these two calves indoors, and only allowed them to suckle, each night and morning.

This was because while they are suckling calves, cows are very shy about their sex life, so now separated for most of the time she was soon mated successfully to an AI bull.

After ten weeks these two were again taken off her, and another pair of calves introduced once again. So by late summer we had seven well reared calves, in a shed to keep for the winter.

That autumn, a neighbour was pleased to purchase this cow with a single calf for a hundred and twelve pounds, so for the cost of a little decent feed, and a little time and patience, I had a little shed, full of nice well grown young bullocks, to sell the following spring.

Yes, there is always something for something, but hardly ever anything for nothing!

Keeping the calves off the cows, and bringing them in to the yard to suckle twice daily was the fool proof method I used to get my suckling cows back in calf regularly and timely, that was up to the time I started, to keep my own Bull.

# Chapter 9

Dabbling with cattle part time.
Buying yearling friesan heifers, and calving them down for sale. Full air for Tomatoes once the Full Moon has past each month of June.

Having jumped on a year or two, let's look back to when I was at a market and I saw a bunch of young bullocks, that I thought, could easily be improved.

When their turn came I bought them, but on the way home I was not completely happy.

Something was niggling at my mind, a kind of unease, I had dosed the six animals for stomach worms, before leaving the market. This was always routine procedure with me, on arriving back home I turned them out onto a small field, and off they went, each one in its own direction as opposed to staying together.

The penny dropped immediately. I had bought a bunch of cattle that had been put together, probably by a dealer. I had always tried to avoid this, as animals that know each other settle down to their new surroundings much quicker. However, here they were, all scattered apart.

That evening I went to have a look at them, and they seemed to be all ok except one, that was lying down with his head flat on the side of the hedge. I slapped his rump, and off he went towards the others. The next morning there he was again, head stretched out on floor. Walking back up I wondered about this animal, and the vicar's cow previously mentioned, came to mind.

I went down again, this time with a piece of bread with me, and sure enough, he was moping again, obviously a spoilt individual. I talked to him and rubbed his back, and he turned his head towards me, and promptly ate the bread I offered him.

Yes, I had a classic loner, that had been coddled.

It took at least three weeks for me to get him weaned off this extra attention. He gradually settled down with his mates, before I sold them on again. Had I not had a little patience with him, I fear that he would have finished up like the vicar's cow.

This proves that animals recognise the way they are treated, and it is up to their owners, to get the best out of them, by treating them humanely at least. This was a good lesson for me, never again did I buy cattle at any market, unless the vendor was present with the auctioneer in the box, at the time of sale.

Early 1966, a five acre smallholding came up for sale, across the road from our nursery, which we managed to buy. This had been a poultry farm for some fourteen years, and had several large poultry sheds around the place.

One of these was gradually adapted for a few cattle, and it was decided to buy a few strong dairy bulling heifers. These were mated, again to AI so by 1967 we were selling a few freshly calved down heifers, at our local cattle market.

Fed well, again using our home mix of good energy food, and well wetted, we found a ready market, for animals that were just a little bit better than average condition.

Prices then were around eighty pounds or so. We achieved a hundred pounds occasionally, and the next year up to a hundred and twenty pounds.

These animals we bought mostly on farm sales around our area. I was now renting bits of land here and there, and I was fully aware of the poor condition that these pastures were in, our most local auctioneer, rather sarcastically dubbed me a hobby farmer, and I soon reminded him that he should show a little respect for his customers!

After all he was charging a hefty commission, on every animal he sold on my behalf.

Next spring, when he called for his dozen or so tomato plants at the nursery, I gave them to him for half price to show no hard feelings.

Early on in July, this gentleman called with me, saying that he was not completely happy with his tomato plants. I suggested he bring one fruit, for me to look at, as I could trust my eyes rather better than my ears, and this he did in a day or two. On examination, it was plain to me that his plants were not getting enough ventilation, as the fruits he brought had halo or round mildew marks on them.

I repeated what I had told all my customers previously, "After mid-June, or certainly after the full moon in June, you must keep some ventilation open in your greenhouse at all times, especially so when growing tomatoes."

He insisted that he opened up his greenhouse early each and every morning, before he went into his office. I replied that the condition of his fruit told me otherwise, so that I could not really believe him. By now quite aggrieved he left, looking back at me in disbelief, he evidently thought that he was being truthful, and so did I.

Ten days afterwards he came in again, saying this time, that I, "The little Jew boy", this was what he used to call me, had been correct all the time.

Apparently, his good lady wife, had been closing the greenhouse, each morning after he had left for his office, thinking that the tomatoes would ripen quicker, in the extra warmth

This he found out when he returned home unexpectedly that morning. Our mutual respect seemed to grow somewhat stronger there and then. Extra heat and sun will never hasten tomatoes to ripen, only time and proper feeding will induce them to ripe fruition.

As well as these dairy heifers, we had one or two nearly finished steers, nearly ready to go as beef. One of these steers took a fancy to a bunch of heifers next door, and was found in the field with them one morning. Our neighbour, asked me respectfully, to please keep him home, I apologised and promised that he would not bother him again, and he did not.

He also got the wire treatment in his ears. Waiting for the correct market, he spent his last fortnight, our side of the boundary hedge, and the neighbour's heifers on the other.

Even so, during this time, this animal lost three quarters of a hundredweight in weight, simply by fretting there, unable to cross the hedge to his new found girlfriends. Even though we had taken the wire off his ears after four days, he could not cross that hedge again.

Animals do suffer psychologically, probably more so than we understand. After all, this animal lost six pounds in weight, each and every day, during those last two weeks!

# Chapter 10

Discovering Potassium Permanganate, the old
fashioned but still the best Antiseptic.

Early 1966 we built our largest green house,
mainly to grow tomatoes to sell to the passing
public. These were a good draw line, that proved to
be very successful.

However, by the middle of the following year we
had a problem developing.

Our tomato plants were dying off, when only half
grown, something was affecting the roots, and as on
farms, anything dead is not of any use.

Our local County Council's horticultural officer
came to look at the problem, and the eventual
diagnosis, was that the water from the well that we
were using contained a nasty little bug known as
Pythium, a form of lethal Phythothera.

This nasty and prolific bug attacks the root
systems of plants and trees, and thrives especially
well in warm conditions, as is found in green houses,
where it breeds prolifically.

The experts at the time recommended we treat
our well water once a week, with Potassium
Permanganate crystals.

The dose needed was only one level teaspoon of
the crystals once a week, for the nominal cost of a
few pence, this was enough to purify the water we
were using, and it certainly put paid to this nasty
bug.

The nursery was using 3,000 gallons of well
water each and every day, and it was incredible that
such a small amount of these crystals was sufficient.
This complete purifier was now present in all our

water at all times, and we used it daily to water all our multitude of plants and trees, to wash our hands, our wellington boots, and everything else, we soon noticed that our hands were cleaner, and that minor cuts healed, much quicker than previously.

This was a break through, that has benefited me and my family every day ever since, as I have also used it for all cleaning purposes, around the farm; when treating the feet of either cattle or sheep, washing down walls, cleaning down concrete yards, even washing our windows, which I do occasionally, without being too fussy! I still to this day, long after retiring, never water anything in my small greenhouse, without adding it to the water, especially for my few tomatoes.

Recently, I was speaking at a packed Hall full of WI members, and one elderly Lady had this to say, "As a pupil in a convent school we girls had to gargle using a weak solution of Pp each morning after breakfast," that is another true example of how good an antiseptic Potassium Permanganate is.

A few years ago I was asked to call with a neighbour, a retired chief engineer, on a ship in the Merchant Navy. He was having trouble with his tomato plants. I told him to call with me, so I could give him some of this water purifier. He looked at me and said, "My dear boy, I have plenty of that. I wash my kitchen surfaces down with it every Friday morning; there is nothing better my boy! We would have damn well lost the second world war without it!"

Apparently during that period he had seen service on a large troop transport ship, carrying troops all over the world, as well as transporting

Italian prisoners of war from Italy to India, as many as a thousand at a time.

He went on to say, "The whole ship was swabbed down daily with Permanganate. Half of those men, would not have survived our voyages without it".

"Yes, damn good stuff my boy!" enough said I think, although I always use it as a gargle, if I feel a sore throat coming on.

This PP is completely safe to use, even to purify drinking water as said, so long as the dilution is right, if the colour is a light pink, all is fine, but if the solution is a dark purple, that is too strong. During recent years our government tried to ban its use, as it is flammable, and could be used to make bombs.

However, by today it is available in good private chemists. If not they can order it for you, but only in 25 gram bottles which is ample for years. Used properly it costs less than a few pence to sterilize a thousand gallons of water.

The larger chain of chemists have told me that it is impossible to get it, simply because large commercial concerns cannot make sufficient profit from handling it. There is absolutely no other excuse. All these fancy disinfectants are nothing more than a commercial rip off, compared to this.

A few years ago I suffered with a number of nasty ulcers on my right leg, these were painful and very slow healing, the nurses were very sympathetic, and truly believed that this was probably going to be my lot for the foreseeable future, and told me so straight up.

I started swabbing my leg with a very weak Potassium Permanganate solution at home and they soon cleared up, I still swab my lower legs twice a week and thankfully, I have had no more of this awful trouble over the latter few years.

Even more recently a Nurseryman told me that he remembers an elderly school nurse using Potassium Permanganate to clear nits in their hair, when she came visiting his school many years ago.

And if that lot of uses is not enough to influence the reader, I recently found out that Potassium Permanganate is even today included as part of survival kits used on long hazardous journeys and expeditions, to be used to purify any water they find, if no clean water is available to drink!

The rule is so long as the treated water is only slightly pink it is safe to drink.

If you cannot find any, try on google, but remember you don't need to buy a large quantity.

A friend of mine bought a three year old bull, of very good breeding, he got him very cheaply as he had a nasty limp in one front foot, this was washed each day with Potassium Permanganate, and soon his foot was put in a bucketful of the solution every morning, in no time the foot was sound as a bell, and this bull has given years of good service since, so never fear using PP to cure and more importantly to avoid foot troubles in cattle and sheep, as well as on wounds of all sorts.

# Chapter 11

Awakening up to the facts of inflation.
Buying a smallholding.

By 1970 inflation was becoming more and more of a concern, and in early spring of 1971 the price of land reached £120 per acre (that is, it had doubled in price during the previous fifteen years). This had never been known to happen before.

Then, in June of that year, I bought a smallholding of sixteen acres, locally reputed to be very good land, for just over £500 per acre. This came with small two bay hay shed and only a derelict small cottage with it.

The sale was held in a local hotel, I had instructed my solicitor, at the last hour on the morning of the sale, to attend, and bid for the little farm for me, not giving him much chance to discuss it with anyone else beforehand.

My instructions to him were precise and clear: he was to be careful, concentrate, and bid only on the odd, and not on even numbers, this was my usual method, with a top limit, of one bid over £8,000.

I was present at the sale and the bidding commenced around £2,000 rising steadily in £200 bids that reached £8,000. An extra bid of £200 was not enough, and the holding was sold for £8,400, so I went home slightly disappointed.

This was an extremely expensive little holding on the day, as I walked up the street along with my next door neighbour, he turned to me saying, "I fear that you are in for some serious competition. Surely it must be a nurseryman that has bought that place for such a high price, he will soon be in competition with you." After some more banter, I arrived home, soon to hear the telephone ringing, and to my surprise it was the solicitor, telling me he had been successful, and that he had bought it for me.

Around six weeks later I was called in to settle the bill, including the expenses. I was now attended to by the head of the practice, he wrote the cheque out to the last penny, and could barely say, thank you to me.

While putting my cheque book back in my pocket, I said, "Now, perhaps we can finish this transaction properly." I pointed out that my instructions on how to bid had been ignored, and that it had cost me the extra £200.

Without a single word, he wrote out a cheque for that amount, and I left for home, quite pleased with £200 in my pocket. A good little win for the small man for once!

The sale had taken place on the 21st of June, and as purchaser, I had immediate access to the land, in order to harvest the grass crop. Prior to the sale the whole season had been extremely wet, and the whole area had grass growing up to three foot six all over, obviously heavily dressed with high Nitrogen fertilizer.

We cleared it all as silage of rather poor quality, just a fill belly really. By mid-August we had bought in twenty eighteen-month old British Friesian bullocks, and by the following spring these were looking well, having been fed all winter, on a good well wetted home mixed feed and the silage as it was.

Another eight weeks or so out on fresh grass, saw them well fattened and ready for market by end of

June, just catching the best market prices for finished beef, that  always dropped during July, these animals brought us a modest profit, all be it on a small scale.

Meanwhile, a decision had been made to prepare half the land to grow small trees for the nursery. This was going to be a costly operation, so it had to be done properly.

Soil tests confirmed my suspicions that only nitrogenous fertilizers had been used for some years. The humus content of the soil was very low, so we kept extra cattle indoors all the year round, using large quantities of straw, and therefore producing a lot of bulky good farmyard manure, relative to the acreage.

Next year we ploughed up four acres to grow potatoes. Having covered the ground with dung all over twice during the winter, this produced a very good crop. After the second year we moved the potatoes on to a fresh patch. By not being greedy, the potatoes paid well, and soon paid for the little farm, as we were retailing the produce ourselves.

We were now planning to start growing trees in 1975, but during 1973, our three local County Councils had a brainwave, they were in future going to buy their trees centrally, all together, from somewhere up in England. We were told that our services, as the local supplier of trees to them, for the last eight years, would not be needed any more. So how lucky we were, that we had delayed planting quite a large number of tree saplings.

We were getting quite used to hard business knocks like this. A few years previously, Harold Wilson as prime minister, stopped all the company perks racket, when all businessmen and top brass could claim Tax back for almost anything against their expenses. This one change alone, reduced our total turnover by nearly 50% at that time as well.

The following year we built ourselves a new cattle shed and we kept 25 bullocks again as before. I also bought 20 weaned calves, from a good local source, and within a week these were all down with a viral Pneumonia. This was totally unexpected and a mystery, until I realised, that I had borrowed a trailer to bring them home, and this proved to be the source of that infection.

Cattle boxes were very few and far between then. However, we managed with care, to save the calves.

That year we also noticed a drop off, in the quality of the Friesian bullocks that we needed, so we decided immediately not to buy any more Friesian bullocks.

Farmers were just then starting to use Holstein bulls on their cows, and passing the resulting calves off as Friesians. The one or two that landed with us were frankly a total dead loss to produce decent beef, as the older they got, the poorer they looked, so that was a quick ending to that job.

That definite decision, not to mess with poor beef cattle, proved to be fundamentally a very good one for us very soon. There is absolutely no single reason for anyone in farming to be afraid to try and keep better animals so long as they are interested in so doing.

Money invested in animals and blood is a much safer bet, to bring back good returns.

Whereas metal and large machinery are necessary, as labour gets more expensive, they have no extra pair of eyes, and they all tend to rust eventually.

# Chapter 12

Deciding, to buy in Better Cattle.
Lucky with calving.
Calves usually bawl a perfect A flat!
Stopping calves scouring for a few pence.
Realising that Bulls are very sensitive.
I never bid on the even at auctions.

A good neighbour had told me once:- "John, only the best cattle are worth keeping, if you are thoroughly committed to do them well, only then is it worthwhile taking them to market, you must always try to sell something that everybody wants, and let them fight each other for them."

So by 1977, I was looking around for some better pedigree cattle. By then there were a few foreign continental breeds around at highly inflated prices.

Importers thought that size was everything, the first lot of imports were rather on the large side and fairly rough over their shoulders, I was aware at the time of serious calving problems with them .

Having said that, my chapter on easy calving explains that wrong diet, and not the size of bulls is the most frequent cause of most calving problems.

My Grandma since ever I could remember, had a picture of a Welsh Black cow hanging on a wall, and this had impressed me very much as a child. This cow always seemed to be looking straight through me, so I bade my time a little to see what turned up.

Sure enough a large herd of pedigree Welsh Blacks came up for sale on the last day of November 1977; a wet drizzly day. I decided to have a look, thinking honestly, that I may be lucky enough perhaps to get a young heifer at least.

98

Unknown to me, earlier that year our government had discontinued paying some subsidy or other to hill and beef farmers, so the large crowd of pedigree breeders present at this sale were reluctant to buy, and the prices were very low indeed, so I bid on a number of cows early on, but did not carry on to buy.

This way the auctioneer, a stranger to me knew instantly that he could not rely on me to carry on bidding, and therefore he would not run me up when I was bidding seriously, I had made up my mind to try and get some heavy in-calf cows, that were coming in later on, according to the catalogue, and by the end of the day, I had bought myself four cows that were due to calve down fairly soon.

When I went to pay at the end of the sale the auctioneer doing his own clerking, was quite haughty with me, his words were, "Well well, you only bought four cows after all that palaver." I decided to bite my lip and ignore him. I was satisfied that I stopped him running me up that day. These in calf cows cost me only 420 guineas each on average.

I was chuffed, and could hardly believe my luck! Collecting them the next day I met the vendor, a lovely man, who was very disappointed with the prices he had received, but he was thankful to me.

"At least you helped things a bit," he said, "they would have gone cheaper still without you." We later became very good friends.

So, next day the 1st of December 1977, those four cows arrived home with us, and luckily very soon we had four newly born calves as well.

These cows had always been kept outside, the south west tip of Pembrokeshire being mostly of a sandy nature was amenable to this way of farming, so coming indoors to these cows meant treatment of some kind or other, and they were not having any of it. It took us some time to get them to feel comfortable when indoors, and this was essential for us on our small acreage.

I soon realised they were very good mothers, giving their all to their offspring. I felt that in general, farmers tended to think that Welsh Blacks could survive on the minimum of keep.

Of course they could, but there is nothing for nothing, but fed decently they were at least at that time capable of competing with the best.

The first cow had calved outside during the night, so next day my youngest son and I managed to get the calf into a shed, but no way would the cow follow the calf indoors, after a while my twelve year old son, a complete stranger to animals, said "Dad, can you make that calf bawl somehow?" I gave the calf a good slap and he bawled quite aloud. "Hah", said my son "That was an A flat, perhaps I can call the cow in", he started mimicking the calf so perfectly that the cow came running in, and when she saw my son she charged straight at him, she had crashed headlong into the gate before my son had landed on the ground on the other side, he looked at me and said "That was too near for comfort, don't ask me to help with the animals ever again" I am convinced that that incident influenced him totally against animals, and he has remained so more or less ever since.

One of these four cows, calved down a very nice heifer calf, born with her front legs bent terribly at the knees. She was actually walking the little she did on her front knees.

I had experienced this condition once before but not so severe. She gradually improved but very slowly, it took five to six weeks or so for her to straighten her legs out properly, by such time she had grown really well, this condition was caused because her bones, were growing faster than the tendons in her legs, so she had to stretch these tendons to allow her legs to straighten.

My Vet, said at the time that this was a very good indication that you have a better than average animal. This was perhaps the first incident, that made me realise, that the first few weeks in a young animal's life, is the most important of all.

This reminded me of my first boss cutting the calves noses on the barbed wire. At this age if you can also get their rumen off to a good start, you will then produce much better stock, at considerable less cost.

While we are thinking of young calves, and the best method of rearing them, long before I retired, we had a set routine, immediately on birth all my calves were given exactly 8cc of good live yoghurt, (the emphasis being on 'live') and as we valued our calves greatly, they would get this exact amount each night, and morning, for ten days, each dose given by an injection syringe, minus the needle, smack into the back of their little mouths.

This was to help strengthen and increase the good bacteria in the gut, but more importantly it helped to establish and keep the correct level of acidity, known as the P H level in their stomachs, so that they could better cope with the quantity of milk they suckled, this also prevents high acidity and any scouring through over feeding.

I believe strongly that a calf that has suffered from scouring, is never ever again as good as one that has not, and they will never be so profitable later in life.

As to the live Yoghurt, we eventually settled on this dose by trial and error; any more will cause them to become constipated, and this must be avoided at all cost.

At a few hours old, we would also inject them with the recommended dose of multi-vitamins, and again at six days old.

This treatment guaranteed that we had no more sleepy and dozy calves, and no more under developed young rumens, resulting in semi potbellied young calves, that are not converting their costly food efficiently.

This extra care proved to be well worth the effort, resulting in heavier and better grown stock later on, an occasional high price for a special animal is always pleasing, but a regular slightly better average price is what brings home the bacon.

Encouraged by our initial success, I now visited the Welsh Black Society sales at Tregaron and Dolgellau, and I gradually managed to buy two or three heavily pregnant heifers. These were now the cream of what was available, and they were costing me usually one or two bids over 600 Guineas. Always trying to bid on the odd, and usually saving a pound or two by doing so each time. Our little herd soon developed slowly and surely.

Meanwhile, as said, we were working hard on improving the structure of our soil, by buying and using plenty of straw, thus producing useful amounts of valuable farmyard manure which was continually improving the soil on our little smallholding.

We were also by then very busy with the landscaping side of the nursery, a business that by now had developed considerably, so our small time farming was mostly confined to what most people would call part time, or hobby farming.

Very recently a teenage friend of mine decided to buy eight cross bred calves to rear on, for the first few weeks he may have been a little ambitious and maybe gave them a little too much milk substitute, to counteract this he religiously gave each calf 8 cc's of fresh live Yoghurt every night and morning for at least the first month to avoid scours, they have also had ample straw as roughage, all this to help the bacteria in their stomachs to develop to their maximum potential, now at six month old these animals are looking fit and healthy, without a trace of belly waste at all, they are converting the food they are given to maximum efficiency, therefore they should turn out a better profit when sold a little further on in age.

# Chapter 12a

Buying a Bull and more
water problems.

Being that we only had a few animals, it was not really economical to keep a bull just for our own use, a local Welsh Black breeder with a very well established small herd was also more or less in the same position, so we agreed in principle to have a look and see if we could fancy and perhaps agree on choosing an animal, that would suit both of us.

So off we went together to a top Welsh Black sale, on arrival we split up, each going his own way to inspect what was on offer, having arranged to meet up later in the canteen to compare notes, there were a good number of particularly good bulls there on the day, and it turned out that both of us only had eyes for one and the same animal.

In due course that afternoon we managed to buy this animal, and I think we both were quite pleased with our work for that day. It was agreed that the bull would stay with me for a month or two, to be fed up and kept growing, he was now put in a good pen with mains water in a drinking bowl.

I took him for regular walks, fed him regularly with well dampened home mixed food, but somehow or other he was not thriving anything near to what he should be, in fact he was hardly maintaining the weight he was on the day we bought him, let alone gaining two kilos a day or so as he should be, according to the feed he was getting.

One evening five weeks on, I rang the fellow we had bought him from, wondering if I could find a clue as to what I was doing wrong if any, we had a long interesting conversation, and it turned out that this animal had never been indoors for one day in his entire life. More importantly he had only ever drunk water from a mountain stream, and we agreed that perhaps this could be the answer.

At nine thirty that evening, I took a three gallon bucket down to our stream, offered it to the bull and blow me down, he eagerly drank it completely dry. Next day we fenced a small paddock securely, and he was turned out here with access to stream water, and thankfully he gained weight each day from then on. We kept this bull for some years, and he developed into a very good example of the Welsh Black breed, he was then sold privately for an useful price, the buyer bought him on condition that his Vet could inspect him for faults, we agreed with this no problem, but when he was examined the Vet found a number of lead pellets in his dewlap, the loose skin hanging from his neck, I gave permission for these to be removed and we never heard no more. A local lad had been having fun taking pot shots at him, if you please.

Bulls are very delicate animals considering their size, the least change can upset them, this one had survived five weeks without enough water, except for the little I had used to dampen his dry feed, and whatever he had licked up when it was raining.

Each time I bought a Bull after that, I made sure that I asked the seller as to what the animal had been eating and drinking. Also each time I sold a Bull I always gave the buyer half a bag of whatever feed the animal was used to, to help him settle down in his new home, what with change of food, water, and home surroundings, I considered all this to be quite upsetting and stressful for an animal, that has previously been so well looked after.

Many years later whilst collecting my morning paper, I often met a prominent local dairy farmer, and one day while we were waiting our turn I casually enquired how things were going with his farming, "Oh very well, really" he replied, "But saying that I bought a Friesian bull six weeks ago and he shows no interest whatsoever in the cows, the Vet has tried everything to no avail, he has even given him injections to increase his libido yesterday, so it is now a case of wait and see."

I then enquired if he was drinking, and he seemed satisfied that he was, saying "Of course he is or he would probably be dead", so that was how we left things as we went our separate ways.

The following week, we met again, and he confirmed that the bull was useless and was to be sold for meat. I again asked him if he was absolutely sure that he was drinking.

This time he looked at me seriously, obviously thinking deeply, and off he went. We met again a few weeks later when he confirmed to me that I had been correct all along and that the bull was now working well!

Another example of how delicate a large animal can be, bulls are also very tender on their back legs, even a single bite off a sheepdog just above his rear hoof is enough to infect the tendons in his whole leg, so never set a dog to move a bull on, as this alone could hasten a premature end for many a bull.

# Chapter 13

Getting my Heifer Noticed,
A health shock for me,
Selling a heifer and buying her back.

Gradually our cattle were improving. The star of our little herd, was the one that had been born with the bent up front legs; She looked a picture, and I wondered what to do with her. I fully realised how lucky I had been, when I purchased her in her mother's tummy. She was now fully registered with the Welsh Black Cattle Society, as bred by me, this only meant, that I owned her when she was born.

The actual breeding was the work of the previous owner, but it was now up to me to do the best I could with her.

Having attended a few W. B. sales, I was getting to know what was going on, within the society, I had noticed that there were a few very well-heeled retired business people, buying animals, early each year, these people always looked after what they bought extraordinarily well, and then they would compete for society prizes, as they sold them on again in the autumn show and sales.

The last thing I wanted was to lose this heifer, but I decided to take a gamble, I entered her for the March sale, and duly presented her to be shown, and sold, at the very top sale, up in North Wales. The catalogue described her correctly, with all her details; her date of birth, to confirm her age, her breeding, and that she was a young empty maiden heifer. This meant that she had not yet been given a chance to breed, which was true.

We took her out to the morning show, in the class, according to age group, and the show steward approached me, saying that my animal was in the wrong class. I begged his pardon, and suggested that he should quickly check the catalogue.

This he did, but he still insisted that he was correct. "In that case," I said, "the only thing you can do is to read her teeth, to prove yourself out of order, that is of course, if you know how to read her teeth? Or better still, we will get two people, and yourself to read her teeth, so that we remove all doubt."

I remember thinking to myself, that perhaps I had been a little cheeky, but I had to stick up for myself, anyhow two of the spectator farmers were asked to examine her teeth, and they both pronounced my heifer to be perfectly correct as described in the sale catalogue, without any doubt.

This man, the show steward did not even bother to look, but remained very quiet and did not seem to be at all happy.

Not having sold an animal in these sales before, we were completely unknown, and we did not expect any favours, and although our entry was head and shoulders ahead of the other entrants in her class, she was placed third, and the numerous spectators around the ring were strangely quiet.

So far so good. I was satisfied that I had stood my ground amid complete strangers, and had been proved at least to be honest. Had the two that had looked at her teeth only hesitated, I would have demanded to have a local Vet to settle the matter.

Later on in the day, considerable interest was shown in the heifer, and in due course, I was waiting my turn to enter the ring to sell.

As I entered the ring it was then that I realised that the auctioneer, was the very same person, as the show steward earlier that morning. "Oh dear," I said to myself, "here we go"

The auctioneer commenced thus, "Well now we move on to the maiden heifer classes, but first of all we have a barren heifer to offer."

This was said just as I came past the box, so I grabbed his microphone, and coolly told the crowded ringside, that I believed, they were not looking for barren heifers in a breeding sale, so therefore I was withdrawing mine from the sale.

Afterwards I soon sold privately, to the very man, I wished to have her, for twenty guineas more than the chosen champion of the day made. Then to be on the safe side of the society rules, I entered this now private sale through the auctioneer's books.

On arrival home, I sent a letter off to the Welsh Black Society secretary, demanding, that he get the Auctioneer to apologise to me for his untoward behaviour.

I soon got a letter back full of excuses, and that locally up there the word barren was always used when referring to larger heifers, and that they had cross checked, this in the best english dictionary, so they were satisfied, with what had been said.

To my mind this was not on, so this devious man got another letter by return, making it amply clear, that he, the Society secretary, and the Auctioneer both, had ten days to send me a full apology, and that this alone would stop legal action.

Both apologies arrived well before the time allowed, without any more to do, along with a promise, never to use the word barren, in a breeding sale again.

According to the dictionary the word 'barren' means 'having failed to breed, after being given the opportunity to do so, as opposed to maiden, that means, not having had the opportunity to breed.'

So now my heifer was on the north wales coast, being prepared for the next autumn shows and sales. I was then a little apprehensive, and wondering what would happen next but still hoping.

The summer of 1980 passed rather uneventfully although we were very busy between everything, what with the nursery and the landscaping.

In early October I suddenly felt quite unwell so I called with my doctor at evening surgery, on my way home from work. He looked at me and said, "It is unusual to see you in our surgery, apart from your escapade with that cow, we have not seen you for nearly twenty years."

I confirmed that I did not feel too well, so he took my blood pressure, and just told me to go home to bed, and stay there, until he came to examine me more fully the next morning.

This I did and late next morning, the doctor arrived with his two partners. They put a heart monitor on me, and asked how I felt. I replied that I did not feel too bad.

They kept talking to each other in subdued tones, asking me if I could understand them. I confirmed that I could, and they just stared at each other saying nothing. Eventually, one of them said that he had never, recorded this level of high blood pressure, from a lucid person before.

He went on, "If we are lucky, within three months, we will find the correct cocktail of drugs to suit you. We will then hopefully have you under control. Whatever happens, I don't think that you will ever be fit to work again. Arrange your affairs and definitely, get rid of your cattle as soon as you can."

Tablets were prescribed and that was that. A very serious bombshell for dear old me. I tried to behave, by taking it easy, and on the first of November my clear out sale saw my few cattle sold, bar two for various reasons. This was actually a very sad day for me, what helped to keep me going was the fact that I could see them being sold. A near neighbour had recently passed away suddenly in his work, with no warning at all. At least I saw my stock selling well.

With one change of tablet, things started to stabilize, but I had a strict warning to take things easy.

I now had two cows, and one fresh bull calf left, and I had also kept back about 400 bales of hay. The autumn sale catalogues were now arriving in the letter box, all members of the Society hadd these,"" and the heifer, that I had sold in March was entered for the North Wales sale in a fortnight.

Hanging around doing nothing was truly alien to me, so I attended the sale where the heifer was now going to be sold, by this time as an in-calf heifer. Her present owner had mated her to a top class bull.

I confidently thought, that things were going smoothly to plan, but when I got there her new owner was present, he had half a dozen or so nice heifers to sell, but alas, not the one that he had bought from me.

When I enquired of the reason for her absence he replied, "Your heifer has turned out to be one of the best of this year's lot, so I have entered her for the Carlisle Show and Sale in a fortnights time".

"I will be disappointed if she does not win champion of the day there." I quietly wished him good luck, and came home wondering, had I over stepped things, I was concerned if my plans were about to go wrong. It was also now, going to be a little more expensive.

Tests were twice weekly carried out on my blood pressure, and thankfully the drugs seemed to be working as required. I asked if I could take a little holiday and the doctor said, "Certainly, take three months off, but to watch myself, and to see a Doctor if I didn't feel up to scratch".

So off to the North of England and the Lake District, my dear wife and I went, on strict conditions that I would not buy back the heifer.

My first boss wife always used to make him promise each time he went out, that he would not 'Burn his fingers', so likewise I also promised.

At least it was an amicable compromise.

We arrived in Carlisle the evening before the sale day. A stroll up to the cattle market confirmed that my heifer was there; safe and looking magnificent. I hardly recognised her myself due to how much she had improved.

The next day she was eventually declared a worthy overall champion of the day, some achievement in a top Scottish sale, and was being much admired. A very confident looking 'Laird' and his lady wife were discussing her finer points and her superb condition.

"She won't look this good for long," he said, "after I turn her out on the hill at home." I queried whether he intended to buy her. He looked down at me saying, "This is the ninth annual Welsh Black sale here at Carlisle, and I have the previous eight champions at home."

As if it was only a matter of time before he had the ninth. I told him that I was her registered breeder, and he conceded, that I had done a good job.

The sale commenced in due course, and when her turn came, I made sure that the said Laird knew that I was bidding, and lo and behold he soon gave up. The hammer fell to me at 620 Guineas if my memory serves me right.

I felt that I had been lucky; I still owned my best ever cow. She had already achieved the highest accolade possible that year, and I was proud of her. Her interim owner had done a rattling good job on her. She was now in-calf to a better bull than I could have had access to, so my gamble had come to fruition. It was now simply a matter of getting her home.

I soon found the vendor, having a meal in the canteen. He offered me some luck money for buying, which I declined saying, "Perhaps you can be so kind as to take her home with you in your lorry to North Wales for me?"

"Of course I can!" he agreed, "And I can do better for you. In ten days time there is the last sale of the year in Aberystwyth. I have my last six heifers to sell there, so it is no problem for me to bring her down as well. How does that suit you?"

I could have hugged that gentleman there and then. How fortunate, I now had no rush or worry about getting her home, this meant no disruption to our few days holiday, a fact that my wife and I much appreciated.

In due course we met up again in Aberystwyth, as arranged, where the heifer was safe and sound, awaiting the last leg of her journey back home.

I insisted that this gentleman and his son, came with me to have dinner, a small price to pay for bringing my cow safely, all the way from Carlisle to Aberystwyth, and I never knew a more amenable retired banker (which is what he was) ever since. I regret to say that I never met the man ever again, more is the pity.

My heifer had arrived at Aberystwyth cattle market some time before I got there that morning, and one very prominent member of the Welsh Black society took a shine to her, enquiring of me regards her breeding, and whether I had any more like her at home,.

I explained what had happened, and that I only had two animals, and this one at the moment, but that one of these was the mother of the one he was looking at, and that she had remained with me after the sale, because she was elderly and nearly blind, and was due to calve soon. "Let me know what happens," he said.

The cow duly had a bull calf and eventually, two years later when the bull was ready, this man bought him from me. He obviously wanted to introduce some of the larger bred South Wales Welsh Black bloodlines into his herd.

He used this bull extensively, and when he had finished using him on his herd, he still kept him, and hired him out all over the place for years afterwards.

The following week, after the sale at Carlisle, the headline in the Scottish Farmers newspaper was "Welsh breeder buys back his own champion of the day". This animal went on to breed me a good calf each year, until she was ten years old, when I sold her for a very good price to another breeder, with her current bull calf at heel, and she went on to breed well for him also.

Occasionally I took this cow out to local shows, where she remained unbeaten. On one occasion she achieved reserve champion to a fine Charolais bull that won overall champion of all beef breeds.

I fully realised that I would never again be so fortunate as to own a cow as good as her. Cows like that are very few and far apart, a once in a lifetime experience, and that with a large dollop of luck as well!

# Chapter 14

### Selling a Home Bred Heifer.

The first pedigree Welsh Black heifer that I bought at Tregaron produced me a heifer calf, and by the time this young calf was two and a half years old, I was fully stocked with animals, so I entered her for the autumn sale, again this was at Tregaron.

By now well on in-calf, she was among a very strong class of five heifers to be shown and judged before the sale.

One of the other competitors in this class, was the very man that had previously sold me the mother of the animal that I was showing that day, and he did all he could, to get between my heifer, and the judge, I had to tell him to move on as I was in no rush, but he continued to do his best to distract the Judges attention, away from my heifer.

This time we had an honest judge and my animal was eventually placed as number one. This gentleman sulked all day. As winner there was a little more cash for me this time.

Before leaving for home, I got an opportunity to tell him how disappointed I was with his bad loser attitude, and boy did he sulk! I will say he did, but fair play, two years later he approached me with hand extended and said, "Sorry"

Having learnt from my third employer when milking all those heifers not to force young heifers too hard, mine were calving first time just before they were three years old, and were fairly well grown by the time they produced their third calf, and usually by the time the fourth calf came along I could afford to sell them, while they were looking at their very best. By then, given average luck, they had already produced me two heifer calves for replacements.

On average, we get bull and heifers usually in equal numbers, and after all, only the best looking young cows sell for the best prices, for the simple reason they should also return a profit for the purchaser, resulting in everybody being happy. In other words a good businessman does his best to give his customers what they are happy to pay for.

Once, while selling a cow at Dolgellau, a shrewd old farmer asked me, "Why are you the only one to sell young cows, just as you have done today?"

I explained to him as above, and said that it was very important to me, farming on a very small scale, not to become over stocked, and that I had to show a modest profit at least. He turned to me and said, "Of all the people here today, I think you are the one that's got it right, and I am very glad I talked to you."

On the way home, mulling over what was said in my mind, I had a strong feeling of satisfaction. That stranger had certainly reassured me.

During 1981 I was still under strict Doctor's orders, not to do any hard work. I now only had the two cows not sold on the clear out sale, plus the one I brought back from Carlisle. By the spring they had all calved. This meant I did not have much that I had to do, but still I had to see to them regularly. I honestly believe that these few animals, truly saved my life, as a sort of occupational therapy.

I certainly had plenty of time, to think back over things, and I did not enjoy twiddling my thumbs. The busy nursery, was certainly out of bounds, so I had plenty of time to think back over my life up to that time.

The smallholding we had bought was rather unique, it was a total of sixteen acres, and consisted of thirteen small fields, the only remaining example in the locality of the olden days, when one acre fields were the norm.

With fields of this size it was simple to keep tabs on workers, as they were expected to mow hay with a scythe, one field, or one acre a day, and these fields were called one day fields. I was sorry to take the hedges down, but with a longish field it was difficult even to turn a tractor and trailer round. We finished up with five reasonably tidy fields that were still small, but much more economical to work in.

# Chapter 14a

Looking Back to using a borrowed horse.

One incident I will never forget, during the summer of 1949 at my first farm, my Boss decided to borrow a horse to help out, over the hay making season.

One mile from our farm there was a very busy abattoir, that only killed horses full time, most of which were imported from Ireland. We were given a choice from about twenty horses out in a field.

The horse we chose led well, and the Boss and I had made up our minds by the time we arrived back home, that he could more or less be trusted. The next day I had the experience of adjusting some old harness to fit him, and after dinner, I was sent to help a neighbour turn a field of hay, this field was on a fairly steep slope, with a quarry half way up on one side, so off we went, with the new horse now in the shafts of a hay swaffer, turning the hay across the top of the field, down the other side and along the bottom, going well, with me walking behind, guiding the horse with long lines, feeling very pleased with him.

This horse had a very tender mouth, and was very easy to drive. We then had to turn left to climb the slope towards the quarry. As soon as we turned uphill this horse stopped, and started backing up until the swaffer wheels hit the hedge. A few strong words, and on he went a yard or so, then back again into the hedge.

This horse was not going to make the little extra effort, to pull the machine up the hill, so our neighbour came down and turned him to go back across the bottom of the field, and he went no problem until the opposite corner. He turned back again towards the opposite corner, and as soon as we turned up hill; he stopped again!   Tom, our neighbour then got his own horse in front harness, and hitched this one in front, so now the front horse pulled our horse and the machine back up to the top of the field.

Unhitching the front horse, Tom said, "Now let me take him. He will either learn quick, or he will end up in a tin, a lot sooner than he thinks!"

Off he went again along the top, and down the other side, but half way down the horse was turned left to come across the middle of the field, as before, being driven with two long lines. They were approaching the top of the quarry and I could not believe, or imagine, what was going to happen. The horse was turned uphill just above the quarry. Again he stopped and started backing!

The neighbour stood well clear as the horse moved backwards, but only, until he felt the extra pressure on his shoulders, when one wheel started to go over the top edge of the quarry, that was the instant he stopped. Though dumb struck, I noticed the horse's ears twitching, as he dug his feet into the dry earth, and made a mighty effort and actually trotted up to the top again, the machine rattling away behind. By then he was in a lather of sweat, induced by the extra adrenaline, and the use of nervous energy, for those few moments he must have been under extreme stress to say the least.

"Now that should teach him!" the neighbour said. "Take him round again the same way so that he turns uphill a few times above the quarry." This I did, and that horse never refused again, we kept him the whole hay making season. When I arrived back home later that afternoon the boss asked me how I felt, saying "You don't look too well, what's wrong?" I must have looked a little pale, after all I had been subjected to a hard lesson that day as well! That horse was definitely no fool.

A previous owner had surely let him have his own way and he (the horse) had taken advantage of that person's weakness, but this animal knew in a split second, when he got into a no win situation, and he immediately recognised the danger he was in, and even struggled desperately for his own preservation.

In fact, I thought that horse was quite a bit smarter than me that day. He turned out to be very easy going, without any other vices at all in the short time I knew and worked with him. Sadly at the end of the harvest season that animal went back, once again to face his inevitable fate.

# Chapter 15

Buying an Improvable Holding and
Getting on with it.
Use Jam jars to assess the quality and
structure of your soil.

Since 1979 my wife and I, now lived in a bungalow, which we had built on the before mentioned smallholding bought in 1971. So we were mostly away from the hurly-burly of our nursery world, which was now being run by our son and his young wife. My life was extremely strange, being not fit and well and in a state of semi idleness, previously unknown to me. I bought one or two animals and did my best to keep thinking in a positive manner.

By mid-1981/82 my health remained constant, so I casually cleared up little jobs that needed finishing, on the landscaping side of the nursery. I was now carefully avoiding stressful situations and by 1982, I thought enough was enough, and decided to look for another smallholding a little larger.

In 1971 land made £500 an acre, but by now land was making £1,500, so had trebled again in a mere 10 years. Looking around there was no land for sale at all, there then came a smallholding, next door but one, to where my father's family had spent their whole married lives.

My Grandfather during his lifetime had quietly, in his spare time, against all odds, transformed what in 1900 had been a rather barren stony mountainside, into a decent parcel of productive land, so I was confident that this land, nearly next door, could not be all bad. This was for sale with a recently built bungalow as well, situated a few hundred yards up the road. On inspecting this land first time, I was ever so shocked at its poor condition.

Hardly anything, had ever been done to this land, for at least a century, the local estate had let this small-holding and the agreements meant that the tenants had to abide with certain restrictions, they were not allowed to touch any game that strayed onto their land, or to take any fish from the boundary stream, likewise, they could only catch one rabbit a week in season, for use at their own dinner table.

Late winter and spring the rabbits were breeding, so no trapping was allowed during that period.

After the Second World War, 1946 or there about, the estate decided to sell this thirty acre or so, smallholding, and a local naval officer who had just retired, purchased it as his retirement home.

He went on to establish a turkey business, and as this developed, the land was mostly used to dump all the manure that his turkey enterprise generated.

Some essential draining work had been done, as there were government grants for work like this after the Second World War, but these drains had not been maintained and by now were in a sorry state, and by the 1980's the fencing posts were all rotted away.

In the early 70's the naval gentleman retired and moved away, and eventually, a local farmer and speculator had purchased the land, without the original farmstead.

He had also bought a bungalow just up the road, to put with the land, ready to sell on, and in the meantime he had let the land off to various neighbours for several years.

No maintenance or improvements had been done to this land at all. It was exactly like moving back at least 100 years in time.

We decided to buy this parcel of land and the bungalow to see what I could do with it. Here I was with 30 acres that included six acres of meadow down the farthest end, the boundary of which was the afore mentioned brook or stream. There was only one entrance into these meadows, a 3-foot metal gate, that animals had to walk through one by one in single-file.

These meadows where covered in rushes, mostly shoulder high, with a patch of grass here and there, and there was no visible wide entrance anywhere to allow a tractor into these fields.

Trash and brambles had by now encroached from all the hedges, for a distance of 10-20 yards. This made it difficult to find if there had ever been another entrance.

After a while the original entrance was found, where by now a row of ash tree seedlings had grown across it, I soon established, that these were at least 20 years old, so more or less no tractor had been in these fields for at least that number of years.

The remaining land was all south facing and well sheltered, again with trash all around the hedges, covering up heaps of stones and boulders that were left along the hedgerows, up to some four yards out into the fields. These had been pulled in there, probably by horses many years ago.

The whole upper fields were under a heavy crop of grass, well ready to be cut, a complete mass of red seed heads, as it only consisted of what we knew of as "Yorkshire Fog", a weedy variety of grass, of very little feed value.

This was in due course cut, and we made it into silage which was bailed and wrapped, our little Dexta tractor could handle these no trouble, as this weed grass weighed very little. They were stacked in the corner of one field and safely fenced off.

I had been told by the vendor of the property, that he had let the land to various local farmers for summer grazing, but in the top fields there were circles of better grasses growing here and there, it was obvious that cattle had been fed during the winter months with good hay, brought in from elsewhere, this had shed its seed whilst being fed, and this grass seed had remained dormant and viable until the correct temperature had come along next spring, when it had germinated successfully, having survived everything the winter weather had exposed it to, this one observation opened out a new improved and better world for me, as I eventually came to understand grasses. I explain all this in the next chapter.

Meanwhile I decided to prioritise by clearing the meadows first, we started mowing the rush exactly in the middle of each of the three fields, going round anti-clockwise, and working out towards the perimeter hedges. Quietly I managed to do it, stopping whenever needed, to cut down tree saplings that were too strong for the mower to cut.

Eventually after some days, I reached the bottom corner, and one afternoon, I noticed two gentlemen walking down towards me, they introduced themselves to me as neighbours, and also as the previous immediate tenants, the eldest went on to congratulate me on the job I was doing, commenting on the bulk of rush and weeds there was.

"I'm not surprised" he said "We did put a lot of artificial fertilizer down here, the last few years."

I replied "Oh!, then perhaps you can explain to me, how you got it in here, and how you spread it, this I am sure, is going to be interesting."

The elderly gentleman looked seriously at the younger and said "Son, I think I have told one lie too many already, this man knows what he is about by the look of things."

"I am surprised that you had not wanted to buy the place yourself." I said, and he replied "I could see it getting too expensive to tidy the place up, as I would like." What he should have said, I am sure was, that he did not fancy the work involved.

During the autumn months I bought about 30 bullocks to keep the winter outside on the driest field available. The silage, such as it was, was quite palatable and it was fed to the cattle behind an electric fence, with some dampened home mixed straight feed as well, this kept them looking decent throughout the winter, but it was obvious to me, that this method, was not an economical way to improve cattle.

The few Welsh Blacks in the bunch, did not seem to out-winter any better than the other mixture of breeds, that I had managed to buy, but on the other hand they were no worse either.

By next spring, cleaning up the top fields, and getting a set of buildings up, became my priority.

The convenient place for building was at the top corner of the land, which was on a slight slope. I got the necessary planning by late spring, and decided to get on with it during the summer.

I booked our local contractors for ten days work, using their largest excavator, also one lorry, and a large dumper, to start work as soon as the ground conditions were good enough, they saw this as a good job at the time, and duly turned up during the first spell of good weather.

The site for the buildings got cleared of topsoil, and this was stored conveniently nearby, we then started clearing the stones off the fields.

The driver of the machine proved to be first class, cool and collected, he did a thorough good job. He steadily reached into the hedgerows and quietly brought out all the brambles, with their roots intact, this was an added bonus, as somebody wilder would have ripped them out leaving the roots, to sprout and grow again. His steady hand reduced my final costs tremendously.

He would then leave this trash in bundles, well out in the field, for me to buck rake away to a convenient fire, which did not go out for the whole week. He also kept the lorry and dumper going, carting the stones onto the new farm building site, that was eventually filled up level.

Tidy building stones were sorted, to be built into the bottom retaining wall. Below the position of this wall, a slurry pit was also dug and this gravelly material was used to blind the larger stones on the new level yard, we were very lucky to hit a deep seam of clay, just at the correct level for the bottom of the pit.

In 8 days, most of the hedges were left clean and tidy, the yard was ready for concrete, the whole job came together tidy, and  completed for the sum of £1,100, this worked out at around £50 an acre, all this and also the levelled yard, and the Pit thrown in.

I was immensely pleased, as by hand the clearing of the brambles alone, would have cost me more than this.

A three-bay corrugated barn 48 feet long was erected on site by early summer, and completed in time, to take next seasons silage. With some casual labour, a lean-to cattle shed was erected on one side before winter.

This was 21-foot wide, half for cubicles, and half as loose housing, the dividing cubicles were erected with a gap, between their front end and the wall, so that the cows could not trap, a young calf and probably damage him.

The opposite side of the shed, a 10-foot lean-to, as a feeding shed, with four cattle pens, off them, so all was getting together, for a decent small cattle unit. The four pens were each capable of holding five large cattle that could be separated so that I was easily in control, of whatever I wanted to do to the animals.

I had bought 4-tons of corrugated plastic covered zinc, for the sheds, came off a temporary umbrella building, used at the steel plant at Port Talbot.
All this came at a very favourable price and was sufficient to clad all the sheds. I then bought one and a half tons of second hand scaffold pipes, and these were used to make all the gates and pens as necessary, with DIY welding.

The gates varied in length from 24-feet down to 12-feet wide according to what was needed, the longest gates with wheels under the opening ends, and a sliding top-hinge on the other.

The following winter I filled the four new cattle pens, each with a different breed, so to simplify comparing the economics of each breed. Pen one had five Welsh Black bullocks. Pen two had five Herefords.

The other pens were all cross-bred continentals, so by careful management, I was able to monitor their performance, the Welsh Blacks turned out to be the most expensive to keep, as they easily ate more than the other breeds, the Herefords ate the least, but tended to carry more fat. It turned out that by far, the most economical were the cross Simmental's, as they grew better and achieved heavier weights than the others, so were clearly better converters of their food, the Limousins were smaller, and tended to gain less weight overall, although, they were the most popular with the butchers at the time.

I visited two farms with Limousins, and they both had handling facilities, which were extremely robust and very high, a sure sign of handling problems, although the cattle themselves seemed very docile, while grazing with their calves.

One small herd I visited were multi-suckling their calves, that is, on approaching them, they called their calves to suckle, like all good sucklers do, and lo and behold, two or three calves suckled each cow, until they had gone round all the cows. I had never seen this done before or ever after, but gradually it was towards the Austrian Simmental breed I was moving.

I managed to buy a very good well grown unregistered Simmental heifer on a weekly mart, she was soon in calf to a Belgian Blue bull. This bull I used was I think the first B B available at the A I centres, I chose this breed on her as a first calver, as these calves tend to be smaller than average when born, compared to the pure simmental, so it was now a game of wait and see.

In the meantime I got the soil on our land analysed for N, P and K, the resulting report came back as a clean sheet with four capital O's on it, this was another first that shocked me at the time, it read Nil Nitrogen (N), Nil Phosphate (P) and Nil Potash (K). Whatever the fourth represented, it was again nil.

The P H reading was slightly under 5, rather on the acid side. This, looked at altogether confirmed that only nitrogen fertilizer had been used by all the tenant farmers over the years.

The impact of all the turkey manure a few years back, had all been taken out, so now I took a few samples to assess the quality of the soil.

This was done simply by filling Jam jars half full of soil, add water to the brim and give them a good shake-up. Left to settle for a few days, the stone will always settle in the bottom, grit next, and then the heavier soils, rough sand will be down among the stones, any lighter important humus will be left on top, or perhaps some will float.   After a few days, if the water is still cloudy, this usually means you have very fine clay. So fine, it takes a long time to settle. Clear water usually confirms there is no fine clay present.

My samples on this occasion confirmed my first suspicions, there was absolutely no essential humus at all present, so poor old me, I had quite a job in front of me to improve this land, to anywhere near the standard I would like it to be. The answer now would be heavy stocking rates, and two full loads of the best straw were ordered immediately.

On the nursery we had made an offset plough, that could plough a furrow outside the tractor wheel. This had been used to produce furrows close along rows of newly erected wooden posts, we could then bury the bottoms of the netting wire to stop rabbits digging their way in under it.

We used this plough to open a furrow in each of the fields and I could only find 3 worms altogether. So put simply, the land was almost completely dead, there were no worms, and no essential humus, just arid and rather acid soil, with hardly any life in it, completely devoid of nutrients, being at the time not much better than a desert.

Obviously, the people that had rented this land had cared for nothing, except for what they could get out of it.

By now I had a slurry pit to take all the effluent off the new yard, but I was not happy with this, so I bought a small second-hand slurry tanker, and regularly, pumped out all the water onto the land, each time without stirring in the solids.

I found that this dirty/clean water did the land a power of good, this did not need much rain to wash it in, and there was no need to keep stock off the fields for long either, the fields all greened up amazingly quick after this thin dressing, proving to me that it must be beneficial more or less immediately for the various bacteria in the soil, this weaker solution is also more suitable for a quick uptake by the grass roots.

Before spreading the water unstirred regularly, I had used a stirrer once, and I did not feel happy with the results. This was how I weighed up the job, I compared spreading slurry to having a cup of tea! Personally I am happy with a rather weak tea, with one spoonful of sugar now and then, but preferably with less sugar, I compare pumping unstirred water to that.

Now well stirred slurry, I compare with a very strong tea with six spoonfuls of sugar and two spoonfuls of salt added as well, I feel that I would not be happy at all having to drink that, in fact I am sure that sort of diet would be very soon harmful to me, likewise spreading well stirred slurry definitely does not keep the worms and beneficial bacteria in any of our soils happy either.

I then had a serious stroke of good luck, one side of my land was on a busy main road, and I had noticed a large machine passing regularly up and down. This machine travelled on four sets of double rubber wheels for roadwork. One day I stopped the driver, who was very amenable, and I gathered he had a regular contract somewhere or other. We soon agreed that once a month or so, he would clean my slurry pit all out, and create a heap of the solid manure on the bank alongside.

Each visit he would turn the by now dry heap, and then add all the fresh stuff out of the pit also. This machine was large enough to reach all that was necessary to clean my little pit completely on each visit.

I fixed a tin box onto a gatepost, and his payment was in there for him, if I happened not to be home when he arrived unannounced. This arrangement was very pleasing to me, my dung heap was now as dry as possible, and was heating up nicely and also breaking down as fast as possible.

I literally felt like a king about this. Why did I say that? I'll tell you, the present gardeners employed by our Queen, turn their compost heaps regularly once a fortnight. Rest assured this is done for a reason, simply to keep things alive and sweet. After this, the contractor spreading for me always commented, that mine was the only farmyard manure he spread, that did not stink, or even smelled strong. The beneficial effect of this system was almost instantaneous and very well worth the little extra time and effort in doing. The regular turning meant the straw was rotting down faster, and the heap was always warm, as it was having some fresh, essential air each time it was turned.

This arrangement, I'm glad to say went on for years, even so I usually had to remind the fellow spreading my dung for me, that I wanted the job done properly and even, from corner to corner, as even spread dung, produced even growth, and better quality crops. I definitely did not want large heaps of manure left here and there over my fields.

Around this time also, I noticed a bunch of eight Welsh Black yearling heifers for sale. They had been reared as single suckling calves. I bought these and looked after them as well as possible. They did quite well for me, when I sold them on 18 months later as in-calf heifers, but the edge was not there. That little something was missing, that makes the difference, this was that extra care during the first few weeks, which pays the dividends, and profits later on, that are very well worth waiting for.

The Simmental cross heifer calved down a male calf, my first Belgian Blue cross calf. This calf was born fairly easily , I cleaned his nasal passages and gave him a good slap, to wake him up, and I was then amazed by his actions, still lying on the floor he obviously noticed his mother's udder, and made a tremendous effort by waddling on his knees, and went straight for a teat. He suckled his tummy full, well before he was able to stand on his feet. I had never seen this done before either. This young heifer proved to be a fantastic mother to her calf and quickly convinced me that the Simmental breed was very well worth looking at. That calf came on extremely well, and achieved reserve, to the supreme in the next years local Christmas fat stock market, this was quite surprising, as I was not one of the in crowd, regularly, attending this market, and certainly I was not, exercising my elbows in the market bar each week. By now I was well aware that the Simmental breed of any decent standard were making absolutely stupid money, averaging to £6,000 for anything that could virtually walk. This was not financially realistic to my way of thinking so I simply put things on hold for a while although I was by now convinced that good cross Simmentals were very good cattle, and fairly cheap to keep and improve, in comparison to other breeds.

At this time we were well settled in on our new little small holding, and I would usually have half a dozen young pigs around the place, bought in as young weaners I would carry them on till they were strong porkers, weighing around 130 lbs dead weight, this meat had been well matured with a strong and tasty bite to it, during the last six weeks of their life they were fed as usual, but I was fortunate in being able to have pallets of over ripe grapes from the wholesale fruit markets, the pigs really enjoyed eating these, and my customers also enjoyed eating the resulting pork, this little special extra flavoured the meat very well, and it really tasted out of this world, just a simple little extra attention to detail that made life and work, so much easier and more interesting.

# Chapter 16

Treating Grass seed the natural and best way.

After clearing the hedges and all of the surface stones initially, I now needed to sow new and therefore more vigorous grass seeds that would provide me with better and more vigorous swards for to graze, or to cut for hay or silage.

My new land was not suitable to plough, due to its stony nature, but some direct seeding had to be done. For many years our nursery had been agents to the main importer of grass seed into the United Kingdom.

At the time, nearly all imported grass seed came into this country through Avon-mouth docks, near Bristol, into a large warehouse, where it was then allocated to all the various seed selling companies. It was economical for us to deal with the main importer directly.

Most grass seed is still imported from California, and New Zealand, where the climate is so much better suited for the job of producing it.

This warehouse was sectioned off, each section of the size to suit each seed selling firm.

Ours was a rather small section, as our nursery sold mainly lawn seeds, with only a few farmers that we knew, ordering what they wanted with us as well.

Grass seed is very resilient, and only if it dries out completely, soon after germination, will you ever lose it, I have personally sown grass seed every month of the year except January; and have yet to get a failure on my hands.

It was usual for me to collect the grass seed I needed direct from this warehouse, each time we needed to stock up, of course these trips coincided with a shopping trip for my wife, thankfully the little lorry we used was quite comfortable.

Many times in dry weather, we have sown grass seed in February, on one occasion I was requested to sow a large area on the 7th of December, this was to complete a large section, when the building of Haverfordwest Hospital was in process.

The main contractor asked me to sow an acre of lawn, so that they could claim a very large stage payment. The weather was favourable, so I sowed half the specified quantity of grass seed. Had it failed, I would have to carry the can and sow it again in the spring, a quite expensive business, but by Christmas the lawn was nice and green.

Grass will always germinate better when sown naturally, that is, simply sow it on the surface and then go away and leave it, as this is how nature does the job.

Mature and established grasses left to their own devices, ripen naturally around the month of August, all good gardening diaries tell gardeners, to prepare for wind damage early to mid-August, it is this wind that distributes the now ripe grass seed naturally. Towards the end of August, the soil is still quite warm, but the nights are getting longer and cooler, and the evening dews are getting heavier, so there is plenty of moisture, an ideal condition for viable grass seed to germinate in.

By mid-September it will be growing well, and it will be strong enough to stand whatever weather the autumn and the following winter throws at it eventually.

It is a complete fallacy to dress grass seed into the soil, as each and every seed stands on one end, before germinating, the new young roots develop all on one end of each and every individual grass seed.

As said this is, the natural and best way, if the reader feels doubtful, all you have to do is fill a shallow tray with any kind of soil, sow a handful of seed on top, water it lightly, and place in light shade. By taking a minute to check daily before it greens up, you will see the seeds all standing on one end.

By today, farmers use contractors to drill in seed with various machines. This, in my opinion, is a complete and utter waste of time and money.

Years ago before we started making Silage, seed used to fall out of ripe hay, and wherever it fell it would germinate, be it on tractor axles, on carts and trailers, and even on narrow lanes as well as on tarmac roads, let alone on concrete yards, so if it falls on soil, or on grassland, it has a much better chance to germinate, should a sheep or cow eat a few seeds, they will survive, and will germinate after the journey through their gut.

So each year during the third week of August, I would broadcast 6 lbs, or 2½ kilos, of the best, and therefore the most expensive late flowering grass seed per acre, the emphasis being on late flowering.

The golden rule to achieve much better results is never to purchase a ready-made mixture.

Oh no. I would always choose the grass mix I wanted, each variety of seed in a separate bag, and then mix them myself to the proper proportions.

Like the cows food, I now knew that I got exactly what I wanted, with no cheap or stale rubbish thrown in, never ever let salesmen convince you otherwise.

I always ordered my grass seed two months in advance, doing it this way gave me time, if there were any hiccups in obtaining supplies, many seed firms and local suppliers only sell ready mixtures, avoid these, as said they may, and probably do, contain a large percentage of very cheap seeds, mainly so that the price seems more attractive to the buyer.

If you want cheap and poor grass fair enough, but if you want worthwhile grass, take your time, either insist well in advance that your local store get you exactly what you want, or find a Firm that will supply you with what you ask for, with no fuss, never leave things to the last minute, buying grass seed is one of the most important purchases you can do, so get it right, after reading carefully a range of seed catalogues.

As said, and I repeat, always buy late flowering varieties, which are rather more expensive, as with animals, only the best is good enough in grass seed.

Late flowering grasses are better because they are always at a younger stage when you decide to cut them for crops, whether you cut early or late, therefore, cheap, early flowering grasses, are always at an older stage whenever you cut them. Younger grasses make better fodder and this is one sure way of harvesting better quality grass crops.

Top sowing at this low rate annually, keeps your grass young and plentiful, as first generation seed, has an amazing amount of extra energy or growing ability, I have never failed with this method over the years, and there was no trace very soon of the old original weedy grasses, that I had inherited when I bought my little bit of land.

I soon noticed the fresh grasses gaining each Spring, they are plain to see, when there is a gentle breeze, and each year they will become more noticeable. Do this diligently you will soon not know what to do with all your grass.

If the reader has any doubt, just do a little strip here and there, and leave the grass itself convince you.

The main enemy of grass seed still in the bag are mice, any seed left hanging about will always attract them even from afar off, the oils in grass seed will soon have their furry coats shining like glass, but the seed itself will be useless after their attention.

I have known a top show man, that bought cheap grass seed and mixed it in with his cattle rations, and just like mice his cattle always had a special bloom or shine on their coats that looked so healthy and natural.

Years ago cattle fed on mature hay always had better looking very shiny coats, due to them consuming the ripe and matured oily seeds heads in the hay, my first employer always swept hay seeds up from the floor, and fed them to his animals.

During the summer of this year 2014 friends of mine have built a new cattle shed, alongside the original farm yard, on the last Sunday of August, their son rang me to say that he was going to sow grass seed all around the building, the weather this year had been particularly dry.

The correct quantity of seed for the area was split into three lots, each lot of seed was sown over the whole area in turns, this way ensures that we get a better even result, with no bare patches, we then walked away and nothing more was done as the soil was quite level and tidy.

On the second day of October, a month on I had a close look at the patch sown and the seed was up strong, even though in the meantime this year we have hardly had any rain, certainly less than half an inch since the sowing, this demonstrates that this is the best time of year to sow grass seed for success, as the dew becomes heavier each evening.

A point that may be of interest is the following, when walking along the roadside, you will find that a tall grass stalk can easily be pulled out of the ground, looking at this you will find that every few inches along the stem there is a node or joint, each one of these nodes if they get in touch with soil will root easily and grow into an identical plant as the one you chose to pull up, or chop this grass up, one inch above and below these nodes, stick these into any soil they will soon make root and each will grow away no problem as a new grass. As said it is easier to grow grass than to kill it.

A few years ago the Welsh Simmental Society held their annual show of Bulls at the Haverfordwest Show. I had a good look at these and I was personally disappointed. That night I rang a friend and asked his opinion, he agreed fully that there was not one animal there with a good, natural shiny coat due to lack of oils in the forage that is now harvested too young. I also noticed that all their feet were rather hot, due to feeding adlib dry food.
Feed cattle, horses, and sheep regularly with well dampened feed, this will avoid you a lot, if not all of the modern foot troubles experienced nowadays.

# Chapter 17

Buying a new breed of Cow.
Easy calving's guaranteed!

I had by now, the mid-1980s, definitely decided to keep Simmental Cattle. Of Austrian origin, they are a larger breed of Beef cow altogether, also capable of producing more milk, a trait that was exactly what I thought I wanted. They had been introduced to this country mainly from Austria and Germany during the early 1970,s and had become very fashionable, mostly in England and Scotland to start with. I made several one day trips to advertised sales, up in the Midlands, whilst there  I got the impression that the farmers selling were not really cattle farmers, they were mostly on very large farms that had possibly joined in the popular rat race, soon after this breed had been introduced to our country

As it was mid-summer, I was amazed how poor the grass swards were generally in Mid-England, compared to what we had in South West Wales, the quality of the cattle offered for sale was also rather disappointing to me, especially the yearlings and young calves.

I felt that the people involved were very half hearted, when it came to caring for cattle, be that as it may, these animals were selling very well indeed, certainly too expensive for me,  a fairly reasonable cow and small calf would easily make £6,000, and I could not, make any economic sense in buying at that sort of price, after all there would be no return on my capital for at least two years, if I bought a cow with a bull calf, and considerably longer, if I bought one with a heifer calf.

After several trips during the summer, I found myself back home, rather disappointed by what I had seen of the cattle husbandry amidst the very large estates or farms employing people to work for them.

By now I had let the Simmental Cattle Society know that I was interested in buying a few animals, so early autumn, a catalogue came through the letter box, of a collective sale in a brand new market, at Ross-on-wye, to be held on the 6th of October.

While looking through this catalogue, I realised that there were quite a few older cows on offer, looking closely, there were two ten year olds from one vendor, these were due to calve before next Christmas, so on the morning of the sale, off I went again quite half-heartedly in the car, not expecting to make much progress at all.

I arrived at the mart ground at 11am, and went round the cattle, carefully checking the catalogue, but the two I had marked out were not present, the sale was soon to start so I watched a few selling, and then I wandered off around the mart again, and surprisingly, the two missing cows were there by now, I enquired as to why he was late, the guy smiled at me and said, that his two young sons did not want him to sell these cows ,so they had hidden them in his neighbours barn that morning, before going to school, hence the delay.

I had a thorough good look at both, each had one tooth missing, otherwise they carried their age very well, their feet and udders were sound, and they both looked heavy in calf,  so at around two-thirty in the afternoon their turn came up to be sold, my first bid secured me the first one, for a thousand and fifty pounds, the auctioneer was a bit greedy, going up in fifty pound bids, but I had once again succeeded while bidding on the odd number, the second cow came in and this one came my way also for Nine hundred and fifty pounds.

I had now bought two animals that needed some extra care, and I soon gathered some straw to put under them, I also bought a small bale of hay, I then borrowed a bucket and gave each a good drink, which they needed badly, after all this I went for my well-earned cup of tea.

With no hope of transport coming down to Wales, there was nothing for it, than to turn the car for home at around four in the afternoon, having first made sure that the market was open till late that night, I arrived home in due course, got into my little lorry, and off I went again to collect them, arriving back at Ross-on-wye Market at five past eight that evening.

My two cows, bless them, were still standing side by side, and it was obvious to me that they had not been off their feet since leaving their old home late that morning.

I had rubber mats in the lorry, and I also put down a good layer of straw, here I was with two old dears, and I was very conscious, that it was up to me to look after them.

After leading them one by one and tying them in the lorry, I parked up out of the way, with no bother or rush at all. Thinking to myself how easy it had been to load these cows, as they were used to being led, a very great asset in my little book, so after a meal in the canteen I got back to the lorry at ten to ten that evening.

My patience had paid off, I was so pleased to see both cows now lying down in the lorry, resting their feet at last, so off I went for home driving quite fast, it was so much easier for me while they were lying on the floor, and they did so all the way home bless them.

I eventually arrived home at one in the morning, having been stopped by the Police doing a routine check just west of Carmarthen.

Back home on the yard, I opened the rear doors of the lorry, took their head collars off, and left them still lying in the lorry, next morning they were eating hay in a rack on the yard real contented, I was pleased how the previous day had turned out, although it had really been a bit of an effort.

It was certainly very pleasing when early December, just eight weeks later, they presented me with a lively heifer calf each, so I was now the proud owner of four pedigree Simmental females.

I could not but help thinking back to the elderly bachelor farmer that I had worked for, exactly like him, I had bought what other farmers did not want, and had succeeded to turn things to my advantage, by using a little care, patience and effort. Of course I also enjoyed that bit of good luck that occasionally makes life so worthwhile.

In January next year a two year old Welsh Black heifer had been brought rather late in the day, into a pedigree sale at Dolgellau.

I happened to see her arriving and I thought that she looked very promising, so I kept an eye on her and eventually bought her, she was leading reasonably well as I put her in my lorry, I eventually arrived home fairly late that evening.

I made up my mind to keep this cow in a pen overnight, while unloading I noticed that she had rather a watery eye, but I never thought no more of it, next morning she was turned out on to the yard and she really looked quite wild, immediately she went and tried to jump over my yard gate, where she got stuck, with her two front feet over the gate, so resting on her chest on the top bar. I was now behind her with a stick stopping her pull herself off the top bar for probably five minutes or so, just to teach her a lesson, she was then put back in the pen, and into the cattle crush, where I tied her ears back over her neck with a piece of wire as mentioned in chapter four,

I now kept her on the yard for four days before I released her ears, then out she went with my cows without no more bother at all, this heifer had been injected to keep her quiet at market, hence the watery eye, and the late arrival at the sale.

Luckily this young cow was soon mated with a good A. I. Bull, and soon she was turning my head each time I saw her, she also seemed to improve better than average. I had decided that I did not want to keep this animal so she was entered to be sold in the next autumn sale which was to be held three weeks after she was due to calve.

However this animal decided to run on a bit and actually calved just ten days before the sale, we experienced a very difficult calving and we were rather lucky to save the calf, this heifer was now back at Dolgellau with her young calf, but showing obvious signs of the difficult calving she had experienced.

Around lunch time a farmer came along and looked at her, and then at me, he smiled and whispered in my ear, "I think you have been a little too mean, a little too mean with the Bran my man".

He went on to explain to me how boiled Bran helps to ensure easy calvings. I thanked him for his advice, and am forever sorry that I did not ask him who he was, as what he told me turned out to be a Godsend to me over and over again, certainly up until the time I finished with farm animals and retired.

The thinking behind this method is to control the amount of nutrients available to the unborn calf, but more important to get the cow to prepare really well and therefore calve down easily.

Around a month before calving, I would start feeding the pregnant cow with Wheat Bran that has been soaked and well mixed with boiling water and then allowed to cool naturally overnight, start introducing this feed gradually until she is eating half a good bucketful twice a day regularly, you will notice that this material when boiled and cooled will be very slippery, so much so that it will slip out through your fingers as you close your hand, this feed has very little food value, but is apparently high in several of the Vitamin B's 1-6, this bulk keeps the cow contented, so not to over eat, the cow will now seem to prepare and be ready to calve possibly up to ten days before the event.

This method guaranteed me an easy calving, was always very worthwhile, and especially when calving down very expensive, or show cows that are perhaps a little too fat, the calf emerges soaking in an oily type of fluid that allows him to slip out so easily, another worthwhile advantage is that there is much less stress on the cow, so she will last a lot longer, the calf will be livelier due to the easier delivery, a sprinkling of salt all over the calf will get the cow licking it more aggressively, a very good point if the weather is very cold, salt also helps induce first time calvers to lick their new calves thoroughly.

Therefore the  benefits of going to all this trouble are definitely very worthwhile, you'll have more live calves, born easily, less strain on the cows that last longer, and go back to the bull quicker, less need for expensive replacements, and less hassle all round, not to mention better profits eventually. When I was on my first farm, he fed his cows very carefully especially before calving, but even his cows did not calve so easily as mine did after this treatment.

With dairy cows their total lactation yields seems very little affected when steaming them for yield is delayed till after calving, and due to the less strain the cows certainly look younger and last longer, this is very worthwhile with all cattle, but especially so with expensive top pedigree animals, and any cow that for any reason is not particularly strong, more effective still if a cow tends to be on the over fat side.

# Chapter 18

Sheep, a new experience.
Buying improvable Ewes and couples.
The importance of clean water for Sheep.

Around this time I was wondering, if I could cope with a few sheep, apart from a fortnight helping a neighbour preparing for his clear-out sale, and turning the handle of the old shearing machines as a child years ago before electrification, I had never been involved with sheep in any shape or form.

By now I had top dressed with good vigorous grass seed successfully for a few seasons, and housing my cattle early each autumn, so to produce more farmyard manure, there was now a fair bit of grass about the place each winter, after all, being south facing and sheltered, in this part of Pembrokeshire, the grass hardly stopped growing even in the winter.

So taking the plunge, I explained my situation to a sheep dealer, and it was agreed, that as I knew absolutely nothing about sheep, I should try a few broken mouthed ewes as a start.

I thought at least they had all lambed before, so they would know more about lambing than me, soon forty five quite strong elderly ewes known as Brokers arrived on my yard, two rams were bought, and things were timed so that lambing was due to start mid-January.

That year a long shed had been erected, to shelter the west side of the farm yard, and the bottom half had been fitted out to suit the needs of the sheep, everything was in place ready, except for myself, I was very aware of an old farmer telling me years ago. "Never you meddle with anything, if you know nothing about it".

So I booked myself in, on a lambing course, with our local Vets, this was to finish just in time before my lot were due to start lambing down, however two ewes lambed with me early January, they had been in lamb when they arrived, luckily there were no complications, and I got away with it, even so I felt completely hopeless, and not at all comfortable.

However, the lambing course was really very efficient and very thorough indeed, and thankfully so, as next week the first lamb to be born turned out to be a twenty four pound monster, the next six ewes presented me with practically every problem highlighted at the lambing course, the remainder thankfully, proved to be very easy, and I soon had a very decent crop of lambs.

I kept them out on grass by day and in the shed at night, the lambs were soon eating a little moistened mixed cereals with their mothers, by mid-March they were off to market as couples, looking well, and turned me a very good profit, and were out of the way well before I wanted to turn the cattle out to grass.

This little episode again, reminded me of the dealer I worked with after I married, these few ewes did well for me. I really enjoyed my experience with the sheep that winter. The next few years I repeated that exercise.

I also bought lambs direct off one farm, some sixty or so in numbers in early autumn, I fattened the male lambs, and they also did me very well, the females I kept and sold them as breeding stock the following year, this farmer seemed very pleased to welcome me back for more, each autumn.

I then decided to buy ewe and lamb couples, the weekly sales for these usually started late January. I would buy only six ewes each with two lambs, once a week, purposely avoiding the best, what I bought were really rather pathetic, and mostly half starved, they were all dosed for worms before I left the market, and I would be back home each week well before mid-day, with six new ewes and their lambs all safely by then in a pen.

After lunch I used a syringe without a needle to give all the lambs exactly two cc's of live Yoghurt down their throat, no less, and certainly no more, again to avoid them becoming constipated, the ewes would then be offered some moistened mixed cereals, and without exception they would take no notice of it.

I would then catch each one in turns, and actually put some dry food into their mouths, and hold it there, by next day they had learnt the score, and were starting to eat, and the lambs would also soon learn to eat with their mothers. The third day they were outside on grass but coming in regularly each evening.

They were soon trusting me, and the mothers were now starting to produce more milk, and for a few days the lambs would get the same dose of Yoghurt, this was to stop them getting scours, due to over acidity in their little stomachs, as said this also helps build up good stomach Bacteria.

Each Monday morning I would now buy in another lot of six, after four weeks, the first lot would go back with me to market, and they never once failed to double the price that I had initially paid for them, this took up a little time, but not a lot else, the feeding was not a large expense, it was surprising how quick the sheep adapted and improved so quickly with that little extra care.

By mid-April I would stop buying, and the last lot would be sold by mid-May. Being careful I, luckily managed to avoid disease, and I had very little foot trouble, mainly because I walked them all regularly, as they came in and out, through a foot bath with a strong solution of the previously mentioned Potassium Permanganate.

The sheep that really stood out for me were the Clun breed, these were outstanding sheep, one of their better traits, was how they guarded their lambs, I once saw three of them obviously co-ordinating together, and chasing a fox hell bent off the premises.

Next autumn, I asked another local dealer to buy me Fifty lambs for to fatten, eventually a lorry turned up around nine o'clock one evening in late September, he had Sixty lambs aboard for me, it was now pitch dark so they were turned out straight from the lorry onto a field where they were eager to graze, next morning they were lying down seemingly contented when I went to see them, so I left them alone without disturbing them, later on I noticed one was limping, I went to have a good look, and found the majority of them were quite lame with foot-rot.

I could well do without this extra work, the footbath was filled with a fairly strong solution of Potassium Permanganate, and before evening they were all walked slowly through this, making sure their feet all had a good soaking, I sent them through this once a day several times and started trimming their feet a few at a time, I had been landed with a really bad lot, but within a fortnight their feet had mostly cleared up remarkably well, and they came on well to be sold fat early next January, that was the last time that I took a delivery of sheep in the dark, as I made sure the dealers knew exactly what was what when I asked them to buy for me after that little episode.

Ever since, I always dabbled with a few sheep, I enjoyed being involved with them, I found sheep to be very honest, always giving of their best to their young, all they needed was a little fair play, one good lesson I learnt was that it is very much well worthwhile, to keep ewes and new born lambs indoors for a few days, until the lambs are nibbling at their mothers dry food.

They will then eat straight away, whenever they have to be brought back indoors again, perhaps due to very bad weather, like young calves, lambs remember their first few days, for people showing sheep leading lambs very young can most certainly be very worthwhile.

To get the best results out of sheep it is important that they have clean water at all times, small tanks are better than large ones whether indoors or outdoors, as they can be cleaned out regularly and often, without wasting too much water, sheep constantly foul their drinking bowls, because they carry bits of food in their mouths, mostly stuck in their teeth, they are very finicky with their water, if this material is allowed to ferment in their water trough even a little, no way will they drink, and their lambs will soon suffer, as clean water is very important to lactating ewes.

I have personally emphasised many times, that the water that sheep are expected to drink, should always be clean enough for their owner to drink as well, if this is so, your sheep will perform to the very best of their ability, their lambs will be ready to sell so much sooner, a little fuss maybe, but all to the owners advantage, this also applies to all farm animals of course.

In the next Chapter I am writing a little about my visits to Austria, and how I benefitted from those experiences, I have also seen our most current and popular television celebrities, both English and Welsh, out on the European Continent, both discussing and marvelling, on how docile the farm animals are out there, compared to ours in this country. I also reveal my experiences after using Trace Elements direct on my grassland, how they changed my whole wellbeing, and indeed my life, let alone the huge benefits to my animals, and to my own pocket! I have purposely not written about the benefit that each individual Element brings with it, as today it is so simple to look these things up, and read about them on the Web, Google etc. and I am sure the truly interested farmer will do so.

Please realise that whatever the problem that arises when dealing with the health of farm animals, the old saying is not far wrong, that is the old saying... The answer lies in the Soil!

When I got my land tested for Trace Elements, the gentleman that did the testing commented on the high rumps and tails on six young heifers I had, he pointed this out to me, and guaranteed that these heifers that were then around six months old, would have perfectly flat straight backs before they reached one  years of age. He explained that this was a spinal deformity due to a chronic lack of Selenium, a vital non-metallic Trace Element.

We went on with a very thorough test of my soil, and my land was treated as recommended, I admit that at the time I thought that the predictions were a little bit too good to be true, however, well before these same heifers were a year old, their backs altered and became as flat as a table without exception.

The gentleman went on to explain to me later on, that we had caught them young enough, that while they were still growing their spines were flexible enough to respond to the better regime of nutrition they were now on after treating the grass as we did.

# Chapter 19

Why were cattle and sheep designed with
such small feet?
Learning whilst on holiday,
Austria's strict quotas on fertilisers,
Testing for Trace elements deficiency.
Black oil fell on my yard.

On one occasion, just as I was finishing speaking
to a Young Farmers Club, I asked them this
question. Why were Cows and Sheep originally
designed with such very small feet, compared to us
humans, when you think of the extra weight, that is
on each foot? The answer is a little further on!

I had already spent one week in Austria looking
very closely at their cattle, and the environment they
enjoyed, and quite frankly I was amazed by what I
found out, bit by bit, whilst I was there.

Firstly, all farm fertilizers are strictly controlled by
quota, the maximum allowed by law was 50kilos per
Hectare (2.2acres).   Their milking cows on the
lowlands were kept indoors at all times, mainly
because land was so expensive compared to ours,
these cows were fed during the summer by zero
grazing methods.

The farms were mostly small family farms, and the
animals were all in superb condition, dry cattle and
young stock were grazed outside on the higher
slopes each year during the summer, and grass up
here was also prolific, and full of wild flowers and
herbs, these cattle also looked superb and were all
extremely docile and quiet, I walked up to quite a
few, and they never bothered even to move, as I
rubbed their backs.

Austria is quite large compared to Wales, and the whole country is above 3,000 ft above sea level, that means it is completely higher than the top of our Mount Snowdon in North Wales.

The atmosphere there is rarer and drier, than ours as they have very little rain, and that only during the summer months, they depend on melted snow from the mountain tops for all their domestic water, and for all other uses, the rivers are crystal clear, being also maintained completely by melting snow.

I have long believed that essential Trace Elements are circulated from the sea, and on to our land in rainwater, if so they are also in snow and stored there until it melts again, so the rivers and the water used by man and beast out there is therefore full of Trace Elements and definitely healthier and better.

Their Dairy products were superb across the board, as far as I was concerned, and I came home very impressed and slightly puzzled. I was also impressed as to how their grass grew, well above the level of the tree line, and equally well each side of most of their mountains, the dark sides of the mountains were lit up enough for the grass to grow by the reflection of the suns light off the ice or snow still left on the top and sides of the next mountain.

I never once saw one bramble or blackberry bush, as they are not able to grow, above an altitude or height of 3,000 ft.

Coming home I felt determined to improve my land and to find out why their cattle were so much more contented and quieter than ours in this country.

Soon after I had a stroke of luck, having made up my mind to buy the best semen that was privately available, I had picked out the bull I wanted to use, and while ordering his frozen semen on the phone I enquired as to how big and heavy this bull was, and whether he was tall or shorter. I was assured as to his size, and the lady I spoke to went on to say that some years ago their cattle had been very much smaller, and were not thriving, so they had called in a specialist to test their land, who had in turn recommended testing their soils for trace elements, this she said had been done by the book, and they had never looked back thereafter, she went on to tell me, that the people that did the testing were Welsh, or at least from South Wales.

This was the final push that I needed, and as said I soon had my land tested by the same people, the analysis revealed that my soil was lacking, to different degrees, in all ten of the most important Trace Elements, namely:

Magnesium, Manganese, Iron, Zinc,
Copper, Molybdenum, Cobalt,
Selenium, Iodine and Boron.

These are all metals, except the last three, but they are all more or less soluble in water.

The total amount needed to put things right in my soil was only a three kilo mixture of all the above, each ingredient measured correctly for each acre, this was to be added, mixed and spread with the fertilizer in early spring.

This took a little time and trouble to do properly, we did the mixing as we filled the drill, and as I sowed, I purposely missed a little area of ground here and there, so that I could compare the result, simply just to satisfy myself, and to expel any doubt or scepticism that may have lingered in my mind at the time, as this was at that time all new science to me.

My cows and sheep were turned out three weeks after this mixture was spread, and only one month later it was really plain to see, that the animals one and all, would not graze the small areas that I had missed out on, the animals one and all knew to the exact inch, what had been treated, and what had not!

Soon both the sheep and cattle were noticeably more contented, they were lying down for longer periods, and my lambs were ready for market, a full month sooner than previously.

The Silage crops were also heavier that year, and all in all, the cost of this initial treatment was well recouped, even in the first year, next winter the cattle were eating considerably less silage, even as much as one third less, this seems rather unbelievable, but certainly quite true, and there were certainly no more pot-bellied animals of any age on my smallholding.

Next winter after treating the grassland I never once heard a single cow grunting whilst they were lying down, this is a sure sign that they are not overeating, and every mouthful they eat more than is necessary, means less profit for their owners!

The following spring at turnout, I was truly amazed as the cows walked on to the grass, they simply started to graze as they got off the concrete, the first ones out would not move on enough to let the ones behind them, get off the yard.

Not one, attempted to run around and kick their heels up, as per usual, this was hard to believe, even when I was there watching them myself.

This alone was a benefit as no fresh grass was trampled on this first turnout each spring.

So next spring I had bales of silage that I could easily afford to sell, having advertised in the local paper, I had quite a hard time getting the price I was asking from an interested farmer, he was rearing young stock for larger farmers, and he needed feed all the year round for young calves.

But the year after, he enquired well in advance if I had any for sale, and all he said was "Keep it for me until I collect it".

Price was not now an issue any more, at least I had one farmer trusting me, he had certainly realised that there was something extra in that silage. Having a little forage over each spring, was a much happier state of affairs than running short of food, this was also one benefit of always making sure, that I was never overstocked.

Next autumn, with plenty of grass about I decided to tack a few ewes. A farmer contacted me, and I agreed to keep thirty ewes from early November till mid-January, this farmer came to inspect his sheep twice during this period, showing more interest and concern, for his animals than most.(with respect).

On one visit I went with him round the sheep, they hardly moved as we approached them from a lower level, as they were all lying down chewing their cuds in bright sunshine.

Marvelling at them being so placid, he took out his mobile phone, and rang his wife, his conversation went like this, "Hello, I am down in Pembrokeshire, looking at the sheep, you know that old "MARY" of a ewe, you won't believe this, I am standing about three meters away from the old devil talking to you, and she is lying down real contented, I would never believe it if I were not here myself".

159

He went on to tell me that he had never been able to get near to that particular ewe, as she would always be the farthest away, and always very difficult to catch.

I explained that it was the Trace Elements that had transformed her, and he was frankly amazed at the result, most especially on that particular ewe. I decided not to tack sheep afterwards, and this fellow seemed genuinely disappointed that I didn't.

I was now beginning to understand the behaviour of my cows, they were now more like their relations in Austria, being much more docile, and they had now tucked up their tummies, so they were carrying much less food in their gut, this meant there was less weight on each of their feet.

A large cow can easily over eat, and is perfectly capable of carrying up to 200lbs of excess wet food in her stomach, she does this subconsciously, because she knows that only by eating more, that she may be lucky to find the essential trace elements that she knows are missing in the diet available to her.

When these are present and correct the cows are contented, therefore they eat less, the result is easily up to 50 lbs less weight on each foot, this is a lot less strain on her feet, therefore the stress and strain on the cow as a whole has been reduced as well, so now satisfied and contented, the cow will lie down for many more hours per day.

This means more profit for the owner, firstly by avoiding seeing to their feet so often, secondly there is less hassle with cows that are not nervous, definitely a lot less repeats, so a tighter calving pattern is easily achieved, also calves are born much easier, all this helps boost profits, simply because you have put the Essential Trace elements into your grass, the natural place where they should be.

Recently a very well-known farmer told me by e-mail that his animals were always fed food that was tested for minerals, and any deficiencies made good at all times, even so his sheep suffered with foot-rot. This gentleman has not yet realised that Trace Elements are over and above, much more important than mere minerals.

Minerals are macro-nutrients, Trace Elements are Micro-Nutriments, and are only needed in very small or minute quantities

A year on after treating my grassland I hired a bull off a large farmer, this bull had been over worked, and needed a long rest, he was to be with me for three months for no charge, so long as he regained weight, with very little work to do he quickly improved, I dosed him for worms and when the time came for him to return home, he was looking very fit and well, but he was not so fat, as his owner expected him to be, simply because he was a satisfied animal, due to the better quality of the grass.

I can honestly say that this treatment of the grass fields, opened out a new life for me,  but I have to say that leading the small calves in their early days  also made life much easier.

Next autumn the representative of the Mineral company called on his annual round, and received an order for only two buckets of supplements, only worth about twenty pounds, he was truly shocked, as previously my annual order was the equivalent of Fifteen Pounds an acre, the cost was now down to around Seventy pence per acre.

I only bought those two buckets just to keep a check, to see if my animals would restart  using them, but no, only new bought ewes coming in for the first few days were interested in them, occasionally, I would see some teeth marks as if a fox or perhaps a badger, had fancied a taste of the salt which was what they were mostly containing anyway, during the winter I would move the buckets from pen to pen on the yard, but they were now completely ignored by the animals.

This new found contentment in the animals, meant  that I needed to use less artificial fertilizers next spring, and the following year I was using only half  the amount I had used previously, before treating my grass, this meant a goodly sum of money in my pocket, and quite a bit less in the merchants pocket.

One afternoon in early March, I was spreading a few loads of dry farmyard manure, and as the sheep were lying down well away from the lane gate, I took a risk and left the gate open, on the opposite side of the lane, my neighbour also had a gate that was open, coming back down in only a few minutes time my sheep were crossing back to my fields, out of the neighbours field, my neighbour had been keeping an eye and could not believe what he saw happening, this reminded me of one of my Grandfathers old sayings, "The best and cheapest way to anchor a sheep is by its tooth" How true, I did not even have to fetch them back!.

Three years on, I was very happy as far as the health and condition of my animals were concerned.

The young calves and lambs were gambolling around like they should be, evidently better for the yoghurt and the multi vitamin treatment during their first few days of life. The better early development of their Rumen (stomachs) resulted in much improved earlier growth, they showed no signs of any pot-bellies, despite the fact that they were getting copious amounts of milk off their mothers.

The cows, and sheep by contrast were as docile as could be, nothing by now would excite them, reminding me again of the cattle out in Austria, quite true, if you show an interest it is from home you learn new things, it seemed to me that all I could do now was to improve the genetics of my stock.

The answer to the question at the beginning of this chapter, why have animals been designed with small feet? My thinking is this, cattle originally evolved with their small feet for one reason, at that time there were many more predators around. Cattle and sheep, are early risers, they are also designed to eat very quickly, then with small feet designed to tire, they then look for shade so to lie down to rest their feet, and to chew their cuds, and by so doing also hiding to a larger degree, from the numerous deadly predators around at the time!

Another point, you rarely see hungry animals lying down contented!! Does all this make sense? That, is up to the reader, one thing that I am sure of is, and here I repeat again, that Sheep and Cows, need long periods each day, to lie down, ideally up to eighteen hours a day.

If they get everything they need in their diet, they will easily achieve this, if their diet is lacking in any essential, they will keep on foraging, looking for that missing element, and therefore by over eating they will carry too much weight in their stomachs and develop unnecessary foot problems, which are caused mainly by being on their feet for too long a period each day, and also by eating too much protein, the resulting bad feet in cows is exactly like gout in human beings, caused by too much Uric acid due to too much protein intake. Likewise sheep develop foot-rot due to deficiencies in their grass causing them to be on their feet for too long.

This first application of Trace Elements to my grassland gave me huge financial benefits all round, let alone the easier life, it was recommended that we tested again after six years just to see whether any top ups would be needed.

Is it worth looking into Trace Elements? Here are two simple home tests you can carry out for yourself, simply rip newspaper into one inch strips and tie onto gates or fences, likewise with any kind of plastic bags or bits of plastic sheeting, your animals will soon investigate these, and will definitely eat them eagerly, as they are capable of knowing what is in them, the newspaper contains Cobalt, and the plastic contains Selenium, two very important Trace elements, the word "Trace" means that only very minute amounts are needed. Another sure sign of deficiencies is seeing your cows browsing off hedges and the lower branches of trees, these contain more Trace elements than grass as their roots are so much deeper in the soil and subsoil.

Another simple and cheap test you can carry out at home, dissolve well by stirring, a small spoonful of Epsom salts in a cup of hot water, add this to two gallons of water and spray this over an area of land that is not too bare of grass, the grass leaves will absorb this mixture much quicker than the grass roots do, within a day or two you should notice your grass greening up, and certainly if animals have access to this patch they will eat it off real bare, confirming to you that they know that this particular grass is improved by what you have done.

These three Trace elements are easily tested for as I have said, simply by taking a little time to do so, and also to observe the reactions of your animals who will tell you that your soil is deficient, you will then need a laboratory test to confirm how deficient these three are, the other seven trace elements before mentioned, will also need to be tested for by a Laboratory, this is important, as you don't need to add too much, you simply need the correct dose, too much is toxic to your animals.

I had also long realised, that our Artificial Insemination organisation, had done stalwart work in improving farm stock generally, they had now been operating since the late nineteen forties, over forty years, but when it came to fine selective breeding, they were not in the frame, the top breeders were not showing their best stock to their buyers, who were buying the best of what they were shown, the old saying comes to mind, "Sell your best, but keep your very best" perhaps truer now than ever.

So with this in mind I decided to clear out all but the best of my cattle, I entered thirty-five of my cattle for sale at our local mart ground, a few across the board from cows and calves to younger stock, well-advertised the sale attracted good interest on the day, farmers questioned me quite closely during the morning, and I felt they realised that the stock presented were honest, at the end of the day we realized well above average weekly mart prices, very respectable indeed, it was plain to me that there was a good market for slightly better than average stock, and that farmers were prepared to pay a little extra for them, but very few it seemed were prepared to go the extra mile and do the work to produce them.

I was pleased not to have any complaints following the sale, and more so when I heard good reports afterwards.

Presentation at sales is very important especially with Bulls, personally I always led my Bulls into the ring, and instead of walking round and round, I would lead them in a figure eight pattern (8) by doing this everyone around the ring had a good view of the animal from both sides, back and front, and all angles.

By watching animals in the ring I always knew what had been well handled or not, animals led round and round on the yard invariably would throw one back leg out wide, this was the one on the outside as he went round, a bad fault, bulls should be led in long straight walks to keep them well balanced, if they are going round in circles, turn them round and go the other way often to keep them balanced.

All my bulls were kept on a very poor feed for a full week before going to market, they were then dunging dry, and therefore easily kept clean at market, I often heard people whispering to each other, saying, "This seems a good bull, but judging by his dung he hasn't had much to eat poor thing, he should be easy to improve if we took him home and fed him a bit better".

Having bought a young bull, it is very worthwhile to take a little time to train him when you start mating him to cows, it is important to lead him out to the first dozen or so, and then take him back immediately after the initial mating, he will soon think that this is how it should be and you will now own a bull that will last so much better, and be much more effective for years longer also.

Here is one memorable instance of how our rainy showers affect our soils. On one Thursday in February 1996 the Sea Empress disaster occurred in our local Milford Haven inland waterway, around seventy tons of crude oil was spilled out into the sea.

On the following Sunday morning the weather was cold and crispy, around eleven o'clock the skies darkened and we had a heavy storm of black/grey hailstones, this lasted for half an hour, I now had a two and a half inches deep cover of this unusual dark stuff on my concrete yard, so out of curiosity I shovelled enough to fill a heaped three gallon bucket, and left it there on the site, later that afternoon all this stuff had melted down to four inches deep in the bottom of the bucket, it was now a deep yellow and black thick gooey oil.

This hopefully once in a lifetime occurrence showed plainly, that our rains do not come from very far out to sea, and I only lived eleven miles away from the source of that storm, that proves our rain is more of a local phenomenon than perhaps we thought before. That is if we thought at all about where or how far our rain came from!!

I was fortunate to meet an amiable farmer in Shropshire, it was autumn time and he had a few young bulls about to come in for the winter, I had a look at these and one nine month old animal stood out for me. I bought him and kept him indoors all winter, doing him as well as I could, by the spring I was pleased with him.

So out I turned him with three heifers, next week he was completely off his four legs, apparently he had damaged a main nerve in his chest while trying to serve a heifer.

This was a very hard learnt lesson, luckily the bull was well insured, so I could afford to go out and buy another.

# Chapter 20

Modest success with Bulls.
Preparing bulls for travelling.
Curing an infectious leg on a young bull.
Selling a bull at the Perth Markets.

I was now concentrating in my small way to improve the stock I had left, and each year I sold a few bulls for breeding, these went far and wide but were mostly sold privately from home.

I would always it seemed have a bull in my lorry wherever I went, this was done just to get them relaxed and used to travelling, even if I only went down to the village, this gradually reduced the stress that young animals suffer, it also reduces the nervous energy they burn off, when placed in a sudden unfamiliar situation. When they are cool and contented, it makes such a big difference to their appearance, when they are presented at market or a show, a blaring radio on the yard, also prepares them for the hustle and bustle, and the extra noise of the market place.

I would take a bull to market, just to get him used to the experience, often the auctioneer would not be best pleased with me, as I often refused to sell, I still feel that auctioneers, don't realise that a good sale is very important to a small farmer, whereas the auctioneer generally tends to be biased towards the buyers, especially dealers, the very people that I was forever trying to avoid.

I have also been known more than once to refuse to sell, just because the dealers were allowed to stand all around the ring, thereby, not allowing people sitting tidily behind them, to see clearly what was being sold, often the Auctioneer would refuse to ask them to sit down when I asked him to do so.

By the time a Bull of mine was ready to be sold, he would be practically bomb proof, well led and handled and well-travelled, this trade peaked for me, one sunny spring day, that proved to be a very lucrative day indeed, I sold three bulls privately at home, two to previous customers, that bought on the telephone without even seeing them, the third one to a brand new first time customer.

Hard work was now paying off, having people trust me was also very reassuring, this alone made life easier and was in itself a very lucrative and pleasing bonus. I would always ask a customer buying a young bull to please give me a first chance to buy the animal back, if and whenever, he would be eventually finished with him.

That year I had one exceptional bull, having advertised him I had a response off a farmer from the Glamorgan area, on seeing him he was pleased to buy him for the asking price of £3,600, he had recently bought one of a close breeding to mine, at the Perth bull sales in Scotland for 6,000 Guineas, and he compared this one favourably, so all was agreed for me to deliver him next week.

The following day, I went to change the ring in his nose for one slightly larger as we had agreed, then I had an awful shock. I had recently repaired the wooden floor boards of the cattle crush, and a piece of board placed across to stop animals slipping had come loose, and the bull stepped fully on to a protruding nail, I was extremely lucky to have noticed what had happened.

I called in the vet and the bull was injected with a strong anti-biotic, as well as this, I was washing his foot twice daily with a strong solution of Potassium Permanganate, that evening I let my customer know what had happened, explaining that there would be at least a months delay before I could be happy to let him go, and fair play, he agreed to wait.

Things turned for the worse and infection set in, he was soon very lame, and not responding to our treatment, eventually an abscess broke out ten inches up his leg, the Vet confirmed that the infection had followed a tendon up his leg, and the inevitable action was to put him down.

"I cannot see him recovering from here", he told me, "You best cut your losses, I shall not call again", this was really devastating for me, not money wise, as he was insured, but I certainly did not like letting a good customer down.

Having a cup of tea later, I now thought the bull was lost anyway, so why not try one more desperate measure. I promptly went down to the local chemist, and bought a bottle of Hydrogen Peroxide, the stuff, when diluted, ladies use to bleach their hair with.

The bulls leg was very warm and painful, and he hardly noticed when I lanced the abscess, a lot of pus was squeezed out, and after washing well, he had 20cc's of neat Hydrogen peroxide injected into the wound, via a Syringe without a needle, this substance was frothing, but the bull did not seem to notice it at all, I repeated this twice a day, using half the strength, and his foot soon cooled down, this was working well, and the poor bull was by now practically lifting his foot for me, and just within a week he was walking without a limp.

A few weeks later he was completely sound and went to his new home with an extra guarantee regards his leg. That bottle of Hydrogen Peroxide was probably the best few pence I ever spent.

The bull never looked back, and all was well when I enquired later. Needless to say that was a lesson well learnt, only bolts were used thereafter in all repairs to the cattle crush floor, definitely no more rushing and using nails in a hurry. Oh No!

Looking forward next spring I was aware that I did not have a Bull suitable, and of the right age, to take to the three local agricultural shows, these were the Cardigan, the Carmarthenshire and the Pembrokeshire shows, so looking around I soon found one in the Cardigan area, he had been done well, all I had to do was to harden him up and give him plenty of exercise, and he developed really well, he won his class in each of the shows, and next autumn I thought that I would try and sell him at the Bull Sales at Perth, up in Scotland.

I had long dreamt of attending these sales, so after making a few phone calls, I soon realised that this was feasible, a long distance haulier based at the village of Winforton, not far from Hay-on-Wye was contacted, and arrangements were made.

I was to take my animal up there the day before, stay B and B with them, then assist the driver on the day, and travel from there up to Perth on the same lorry.

Arriving at their depot at four pm, I soon bedded my bull down for the night, and then helped get the lorry ready. I had two rubber mats which I put in the front of the lorry ready for the morning, and this was the first time that these drivers had come across these thick mats.

5.30am next morning off we went, picking up a bull here and there, eventually there were ten bulls in the lorry, each one in his own compartment, that was lined with a tarpaulin type of material, to avoid them rubbing or damaging themselves, around eleven o'clock we were finally fully loaded and were now on the motorway heading north.

Stopping at Carlisle market, which is near the motorway, for a meal, we noticed that my bull was the only one in the lorry lying down, and so he was when we started off again, we finally arrived at Perth market at 6.30pm, and my bull was still lying down.

While washing down the lorry afterwards the driver told all the other drivers about the mats, and how the bull had laid down nearly all the way, it seemed that rubber mats were new to them also.

That evening, I weighed my bull, and he had only lost twelve pounds in weight during his journey, this was exceptional, as usually a loss of around forty to fifty kilos was the norm for a journey this long, his training meant that he was properly relaxed travelling, but the mats also helped greatly.

This was on a Saturday evening and by Monday morning he had regained the little weight he had lost.

The Simmental sale was to be held the following Wednesday, I had arranged a taxi to ferry me three miles or so to the nearest available B and B, and also to collect me each morning at 5.30, and again to take me back each night at 9.30pm.

These few days at this market were a real eye opener, or perhaps better described as an extra education, established herds, had people there preparing their stalls for them, by laying straw down up to three feet deep, then tramping it down solid and repeating the process several times, straw was plentiful, and was supplied by the local council, who had four men carting it out of a nearby shed using five trailers. They would leave a full load on each corner of the market in turns, and by the time they were back round, there was always an empty trailer waiting for them, this went on from 6am to 6pm each day for a week, the duration of this particular market.

I remember calculating that the council, sold £35,000 worth of straw at this one weekly sale alone, this was possible to do, as they charged a set fee for straw for each animal present, all this regardless of how much you used.

As said, The Simmental breed was to arrive on the Saturday as we did, and were to be sold on the following Wednesday, but on Sunday morning there were four gaps around where my bull was stalled.

The Irish bulls were delayed due to rough weather out on the Irish sea, arriving at 11.30am these bulls were in a sorry state due to sea sickness, and they were soon laying down completely oblivious to their surroundings and really ill.

Their owners were Irish farmers, who brought their children with them, and being late, there was no time for food. As my bull was ready, and nothing more to be done, I went down to the mart canteen and bought ten plates, full of mashed potatoes mixed with turnips and kibbled corn beef, with a few extra plates, bringing these back I said,

"You have time to gobble this up surely," they were so grateful that they saw to it that my food was paid for up until I left for home on the next Wednesday evening, fair play.

These people were all very nice and friendly, down to earth people, obviously very well to do, but they trained their children to accept responsibilities, on the Tuesday afternoon while the Simmental females were being sold, they even got the children bidding for the ones they fancied.

One thirteen year old boy helped me, my bull had decided to drag his heels a bit, so I had this lad to follow behind with a stick, just to keep him going as I led him for a quick walk around the perimeter of the market early each morning and evening, it was important not to let the bull stiffen up, as a bull looks fitter, if he is bouncy when in the ring to be sold.

For this help I gave the lad 50 pence each time. Later on I asked him what would he do with this money, his reply was "Well, I would very much like to buy a Black Moil cow when I can, this money will go in the pot for now, and we shall see what happens when I have enough, as dad is not happy with the idea just now".

That Sunday there were a lot of people inspecting the bulls, then on Monday each one in their turn, had to get past the Society inspectors, and I fear that a few failed, and had to be withdrawn.

On Tuesday they were judged, what an unique experience, there I was, in the ring together with another 28 large bulls, all in the same class.

I felt like a Cork on the open sea, being pushed around, as people jostled for the best positions.

The judge, a rather new member of the Society, having spent a lot of money whilst establishing his herd, and thereby achieving some prominence, was clearly, in my opinion at least, well out of his depth, he seemed overwhelmed with the task in front of him, eventually, he placed around half the animals in some sort of order, then moved them around a bit, with no sort of confidence at all.

I was fairly well placed amid the top half of the class which was quite respectable, considering the numbers involved, and the exceptional quality of all the stock.

Next morning on the Wednesday my bull was sold quite respectably, and found a new home down in Cornwall, I could have achieved the price easily at home, but the experience of meeting farmers from all over the globe practically, was a very worthwhile lesson, in the art of different styles of living, all within the same basic agriculturally dependent world, I only wish that I could have had this opportunity much earlier in my life, though late for me as it was, I benefited tremendously, if only in a boost of confidence.

The experience of waiting five whole days just to sell one bull was unique, each stage of the process negotiated was a lesson learnt, seeing the intense dedication of the top stock men alone, was really incredible, the huge Market area was a buzz of commerce of all descriptions, let alone the new realisation of how tough it was, at the time, for all the small farmers and crofters, living on the numerous offshore islands all around Scotland.

These people have to produce top quality cattle to take to market, simply because, of the extra cost of shipping their animals to the mainland to sell.

I am sure that I, and most farmers would up the ante, and think twice, if they had to pay an extra £120 to take each animal to market, as this was the cost per animal for using the ferries from the Islands to the mainland in the mid-nineties, let alone today,

That afternoon at 4.30pm we set off back for home, with the empty lorry bombing down the motorways, I arrived home in Pembrokeshire a satisfied, tired, and perhaps a wiser man at 2:30 am next morning.

Now in my sixties, with my health fully restored I was contented to keep a few cows, but I was always anxious to do a little better, so I carried on quietly, and when retirement age arrived I gave in to father time, I let part of my land as a summer let for one season, and decided that this was definitely not the best way forward, and made up my mind to sell the little holding next year. I sold my cows as they calved, my working life had slipped away very quickly looking back at it, and I was now seriously thinking of selling my smallholding.

# Chapter 21

Dealing with a Better dressed man.
Avoid the Smells that attract unwelcome insects.
Why spray Sheep with a Soap and Paraffin mixture.
Retirement and Money Saving Tips.


During the late spring of 1998 I asked a local Estate Agent, to look over my place for to value it, and he arranged to see me at 2.30pm on a set day, he eventually arrived at 5.20pm, announcing that he was in a hurry, this promptly reminded me of my grandfather once again, he would say,

"If you are in a hurry today, you will not be at your best I fear, so leave it till tomorrow and you will do a better job of it then, young lad."

So, I promptly told this important gentleman, in a hurry, that we better postpone things on that day, and that I would give him one more chance on a day to suit him, when perhaps he would be in less of a hurry, and possibly be on time, he looked at me rather taken aback, and rather quietly went away.

When he returned he was rather subdued, and I was not overly impressed with his attitude, I could somehow sense his lack of interest in numerous details, his valuation seemed to be on acreage alone, and after some thought and knowing the trends at the time I mentioned to one or two of the locals, that I was thinking of retiring and selling, and I soon had a reasonable offer for the holding from a local young couple, so my smallholding was soon sold privately.

The steady work in improving the holding, and especially the improved fertility of the soil, had paid off quite well, there were now  only three cows left on the place, and the new landowner bought these as well, one was due to calve down in only a week, so she was sold as guaranteed in calf.

This cow, a white Charolais, showed very little signs of being pregnant at all, she had always calved down before developing an udder, but on the day she was due I visited, and there she was proud as brass with her calf, whether this is a breed trait I do not know, but this was her way of doing things.

My wife and I, now retired to our new home where I settled down to do a little bit of gardening, using the little knowledge that my years on the nursery had brought me, I showed a few vegetables here and there, without over doing things.

I had long been of the view that anybody who dominated the summer vegetable shows, by winning across the board, did as much harm to those local shows as good. Given half a chance by winning something small here and there, people new to gardening and showing, can be encouraged to compete, and perhaps progress, to the benefit of these active communities, and especially the volunteers who work so really hard each year.

After retiring, I sold a wood splitting machine to a local hill farmer, and we became good friends. I soon realised that as well as a good machinery man, he was also naturally, an exceptional stock man, and gradually I managed to convince him that he should try to improve his land.

He started by top sowing a few acres with good grass seed in late August as I had done previously. He soon recognised the benefits, and the following year he decided to have a full Trace Element test of his silage fields. A little later, a soil sample was carefully taken for analysis off forty acres of his silage land, these were sent off properly labelled.

The results came back not too bad, low in some things, and better in others, and certainly much better than my results a few years ago on my little place.

The prescribed treatment was carried out, and the following year the rest of his land was also treated. Two years later on, this farmer confirmed that he was now using considerably less inputs compared to what he had been using before testing.

He also confirmed that his lambs were easily finishing one whole month earlier than they did the previous years, and that his life was rather a whole lot easier all round.

Later, he started keeping better cattle and by now they are coming on in leaps and bounds, and I am ever so pleased for him.

Fair play, he was only reaping what he deserved; a just reward for showing a little more care and attention to detail.

One sunny afternoon soon after retiring, I found myself on top of the roof of my new home inspecting a chimney, I was also enjoying the sunshine, I suddenly became aware of the much stronger scents at this height, compared to scents at ground level.

This may be due to the warm air, but there again we don't see many flying insects in cloudy or cold weather. I was now about twenty feet or so directly above a small bed of mixed herbs that had been planted the previous week, and I was really surprised when I realized that I could easily pick out, and identify, each and every one of the few varieties there were, quite separately.

I pondered about this and suddenly realised, that using their keen sense of smell is what enables the bees, and various insects, to find their way to their favourite plants. Their uncanny ability to choose a single plant among a plethora of others is quite frankly remarkable, and, by so doing, they find the nectar they need and also naturally pollinate the chosen plant at the same time.

I think I am correct in saying that, good, and bad smells, dissipate upwards naturally, and become much stronger the higher they get in the probably rarer atmosphere.

I am convinced that all our numerous insects are drawn in like this, the bad ones bring in new diseases that we do not want or need, and the beneficial insects certainly do most, if not all the pollination that is naturally vital to all plant life.

This tells me that it is up to the livestock farmers to be careful to avoid having bad smells around their farms, thereby removing the attraction that brings these evil insects in to plague him.

While sitting on the crest of that roof, my thoughts went back to my first employer. He kept a leaky drum of strong disinfectant constantly in his cowshed, and also a large dab of smelly Stockholm tar under the shafts of the market trap, these were meant, and did repel flies. Again, when a cow lost one teat completely at the next farm I worked at, Stockholm tar was used, all to do with scents and smells used to avoid flies and further trouble.

I later discussed my thoughts on smells with my farmer friend. It was decided on a very hot day to spray the back ends of his sheep with a mixture of soapy water and paraffin, a half and half mixture, hoping that this would perhaps confuse the cruel flies that strike sheep and lay their eggs on or near their feet, resulting in maggots and a lot of extra work, let alone the suffering for the poor sheep. This first time experiment was done a few weeks after shearing.

Next day, the effect of the paraffin was plain from the next field, as opposed to the usual smells of sheep, we were very pleased with this result as he had no more fly strikes again that year. The treatment that cost only a few pennies, was repeated, when he thought the paraffin smell was wearing off, and getting rather less effective, later on a row of old hessian sacks were hanged over a rope between two gate posts, these sacks were sprayed with the same mixture of soapy water and paraffin, and they were positioned at a height so they just touched the backs of the sheep, after being driven under these once or twice the sheep accepted the presence of the sacks and soon took no notice of them at all, it was now easy just to spray the sacks as necessary and the sheep were all protected against fly strike with very little cost and minimal labour.

A few years later, I came across another example that left no doubt in my mind of the effectiveness of confusing flies and insects including the Cabbage White Butterflies. It had been a very wet summer's morning, but later it came out sunny and warm.

A teenager grandson of mine and I went down to my vegetable plot, and as we approached, we both noticed nine White Cabbage butterflies hovering over my few cabbages. As we got nearer, they disappeared over the roof of a nearby house, obviously making use of a thermal current. Nearby, a wheelbarrow was quarter full of clean rain water.

The remains of a few lettuce plants were chopped up finely and put into this water, and was very well stirred. This now yellowy soup was strained into a watering can, and in just a few minutes my cabbages were sprayed all over with this mixture.

My grandson by now was convinced that I was definitely losing a marble or two, and telling me so, especially when we retreated about fifty yards and sat down to wait. Within a few minutes, the White Butterflies returned exactly the same route that they had gone by. They hovered around above the cabbages for a few minutes, and eventually flew off over the house once again, the strange smell of the lettuce tea was enough to confuse them completely.

They could not pick up the scent of the cabbages as they had done before we sprayed, and we did not see them again that afternoon.

I would not expect this to be effective against the white butterfly for long, especially after rain, but the scented oil treatment on cows and sheep is definitely very worthwhile. Also, a rag moistened with paraffin is very efficient in keeping lice off cattle during the long indoor period of late winter and spring.

These lice cause great discomfort to cattle of all ages during this time, especially before turnout, the animals are so stressed they rub their coats bare in patches. Cattle presented to market in this condition, looking rough, are just what the dealers want as they know a few April showers soon gets rid of these lice. The animals then soon look much improved and therefore more profitable for them.

I have personally heard these dealers grumbling at the marts when cattle are presented for sale in early spring looking tidy clean and well, saying that they are too good and will take too long to improve.

What absolute nonsense! All the dealers want is a quick buck at the expense of the farmer.

A year or so on, my theory as to insects depending on stronger scents to guide them was confirmed by our farmer friend, namely Mr Adam Henson, the ever popular star of the television Country File programme. This gentleman once took a recently retired dairy farmer to look over a new super dairy complex somewhere in America.

This place had many thousands of cows, and as they approached in a helicopter, Mr Adams' words were to this effect: "The whole place looks very big and very impressive from the air, but by golly it does not half stink!" This, or words to that effect, were said from inside a helicopter whilst approaching the massive complex, confirming perfectly my humble thoughts that the higher up in the air you are, the stronger the smells become. Likewise with spraying, whilst aerial spraying a fraction of the quantity is more effective than when we spray at ground level.

By complete coincidence, as I write this in early 2013, the awful news of the Smallenburgh Virus affecting new born lambs has just become publicly known. At this time, it has arrived mostly on the coastal areas nearest the European continent. That is not far away when you consider the speed that insects and bees can travel on a following wind.

As a mere clod hopper, I am given to understand that these insects, just like the more common greenfly we all know, are even equipped with more than one nose to help them detect their prey and therefore survive spray attacks.

Surely the government should immediately prioritise our entomologists to tackle this situation before it gets out of hand. I think that it is also up to each and every farmer to do his bit to protect his valuable stock. It is in their own interests to keep their farm surrounds and their stock smelling as sweetly as possible. Planting strong smelling herbs such as wormwood etc. can surely help in avoiding the problems that these bugs bring with them, let alone the financial benefits of avoiding trouble..

Since the development of slurry pits to store effluent, another disadvantage has been the way that Dock weeds have increased. As with all seed, the Dock seed is also very versatile. They are much alike our grass seed: virtually indestructible.

They can survive all conditions except too high a temperature. The old dung heaps of the past warmed up hotter than we thought and killed them alongside most other weed seeds. The Dock plant consists of a crown from which the leaves sprout fresh each spring, and of course so do their roots. Remove this crown part and the roots left behind will not sprout, but will rot and die. True, if you split the crown, both pieces will sprout afresh. A sharp spade will easily remove the crown without going very deep. Like picking stones, you only have to do it once. By now, there are strong sprays that will kill docks, fair enough, but what else do they kill? They definitely kill earthworms to say the least.

These sprays have been developed from what we knew as Paraquat, or Weedol, for at least the past forty years. This stuff can also make one very depressed, even if it is still in its container or box in your office. One boffin working in Porta-down has been known to say that contamination on your skin today could very likely affect your offspring and even your grandchildren, two generations later even, perish the thought!

What we do not realise is that this spray material was first used in the First World War as gas. The Germans ceased using it after the wind turned and blew it back towards themselves, killing thousands of their own men in one day. And here we are using basically the same terrible material to save a few pence and perhaps a little time.

My personal involvement with farming started in the very early years of my life. As a schoolboy, mother encouraged me and my brothers in this direction because at the young age of seven, she suffered the terrible experience of seeing an older brother discharged from the British Army as being disabled. He was suffering from shell shock and had lost the use of most of his lungs due to the very same first world wartime gas attacks in France.

My mother and her sister were brought up in a household that were left by themselves to try and cope with the nursing of this poor fellow, hanging on to life with the merest of threads for several years before he eventually died.

I now realise that it was little wonder why she was so determined and passionately against her sons joining the army as conscripts later on in her life, even though it was then peacetime and well after the Second World War.

By now, at eighty years of age, I thoroughly enjoy my life as each day, season and year goes by. I enjoy a little bit of gardening, mostly in the spring, and also, of course, to a lesser degree, all the year round.

My wife and I also enjoy eating the organic produce we get by doing so. I also watch the birds come and go to the feeder, and the trees coming to leaf each spring.

Talking of trees, I think I was the first to notice the Dutch Elm disease! it first established itself in our country on the very western tip of South West Pembrokeshire, mid 1970s. It then spread like wild fire throughout Wales and then across to England until it decimated the whole Elm tree population without exception.

Today, nearly forty years on, I see fresh young seedling Elms doing their best each year to re-establish themselves in our hedgerows, but to no avail. Each year by August, their young leaves have been ravished all over again by that terrible bug, it is rather unbelievable that Elm seeds have remained viable in the hedgerows for all this time.

And now, in 2013, the Ash tree is again subjected to or is a victim of another disease so foolishly imported into this island. This was absolutely and totally unnecessary as each and every native Ash tree in our country produces thousands of viable seedlings every year. These could easily have been collected and grown on in this country as I did myself on a small scale during the 1960s. I personally can only hope that this fungal disease will not completely wipe the Ash out like the Elm.

Every June I can be seen collecting some Elder flowers. These I trim and pack into the freezer. They make me a tea that has proved to be an excellent diuretic that keeps my prostate problems at bay, at least it has done so for me.

It is also a guarantee against coughs and colds when they threaten. From July on I am harvesting my garden produce, and I certainly know the advantage of doing this. Either to enjoy on my plate the same day, or to store in the freezers till next winter, then during August, I am collecting Blackberries, again for the fridge freezer. September sees me looking for a few Hazelnuts, something I have done each year for as long as I can remember, but alas these last few years there have been only a few mature and ripe hazel nuts in our locality.

I strongly believe that wild fruits found growing on hedgerows contains more vitamin C, this vitamin enables our bodies to make more of the essential vitamin B12 that is so important to our health.

In recent years I have been aware of half ripe nuts that are to be seen all over the floor. These have been plundered and rendered useless, well before they are properly ripe and viable by another import to our country, that we could well have done without.

This is the Grey American Squirrel by now endemic in our country. As sure as eggs are eggs, if these are not controlled, very soon, this country will witness the demise of our native Hazelnut trees as well. If no ripe nuts fall to earth to germinate naturally, this vital shrubby tree will also die out. More so, as nut trees are not so long lived as our larger trees.

As a child, I remember vying with our Red Squirrel for a few nuts. He would know exactly when the nuts were fully ripe, and if we left it a week later before fetching our share, the squirrel would have had the whole lot stored away for the winter and we, as children, would accept defeat without any malice.

I remember clearly as a nine year old, when a Red squirrel stole my first ever so precious ripe Strawberry. I had accused my younger brother of pinching this first one that had ripened in the newly planted bed. On his denial, I, on my mother's stern instructions, settled down to watch out for the culprit. Sure enough, before too long, who popped in through the hedge but a Red Squirrel!

He actually found another ripe fruit, right before my very eyes, and I swear that I saw a sly smile on his face, as he looked back at me as he jumped out through the boundary hedge on his hind legs, carrying my precious fruit in his front paws.

I will never forget that incident. My mother, bless her, quite rightly made me say sorry to my younger brother for wrongly accusing him of pinching. To this day, I can feel a damp patch on my trousers. I must have been sitting on some wet grass as I watched out for that thief.

There were definitely no Grey Squirrels around our area when I was a boy, yet by now, this imported devil has seen off most of our native reds, bar for a few on Anglesey and a few more in Scotland.

By today, I am more than a little concerned as to the future of the younger members of my family as part of the mixed human race as it is today on our island.

That is more or less how I have enjoyed more than a decade of my life since retiring, but when the weather is too wet or miserable, I can usually be found with a file or rasp in my hand, trying to shape myself a better walking stick which, alas, I know that soon I will be needing its assistance.

I am convinced and happy to use one occasionally, and have recently found out that one can be a very faithful and useful friend indeed.

Occasionally my mind wanders and I remember reading the history of a century or so ago, and back to the 1880s, following the successful development of freezer ships, soon after the New Zealanders were dumping their frozen lamb on our island, this caused a serious recession at the time for our farmers.

By today, in 2013 these people of the southern Hemisphere have lucrative markets in the Far East, and it is not impossible that this trend will develop more so, as other countries are becoming more affluent as well.

I feel, research is badly needed into why farm animals, and the land they graze on, are so symbiotic. I am sure farm animals could become considerably more profitable, if and when their owners will take a little of their time to study, and understand their land, and the soils that sustains them both, certainly this will become even more important, as our population increases.

I can foresee food shortages looming sooner or later. Farmers, as prime producers of food need to realise that with care, costs can be greatly reduced and/or avoided, as they are constantly planning ahead and investing in their individual enterprises, I can only hope that one or two of my personal experiences may help to some degree towards any eventual better profits for my readers one and all.

In chapter Five I mentioned the academic newcomer to farming, he held a PhD degree, and he emphasised the importance of roughage for his milking cows, this is equally important for Sucklers as well, in the month of October a Cow suckling a rather large calf is under a fair bit of stress, a cold night can easily tip the balance and you have a cow down with staggers, a mouthful or two of dry hay will go a long way to avoid this loss as dry hay has a fair bit of Magnesium in it.

If in October you have a rough field with dried out thistles not yet been cut, turn some cows in and watch them eagerly eating those thistles clean as well as the grass.

Fifty odd years ago my brother was the Cow man at a prominent Pembrokeshire dairy farm, one day his boss called in with me and told me, that my brother had given him an ultimatum, he had told him, "That the straw rick was now very low, if the straw runs out, that is the day I go up the road as well." He believed passionately in keeping straw in front of the cows every day of the year. And his boss admitted that he produced more milk for him than he had ever seen before.

A neighbour recently told me that his suckler cows were refusing to eat good quality silage that was offered to them, this was in late November, whilst the cattle were kept on some rough over grown grassland while waiting to be brought in for the winter, these animals were contented and happy living on dead grass and weeds, rather than a higher protein diet offered in the silage, roughage is essential, and your cows can speak volumes if you listen and observe them.

# Chapter 22

Report to my local Hon M.P. after testing soils off six various farms for their ability to absorb and hold rainwater, the capability of soils to hold moisture is the most essential part of rich productive farm soils.

24/9/14

Dear Sir,

Last February I sent you a letter regards the serious flooding experienced in England during last winter, your kind reply invited me to try and give you evidence, the following report is the result of some research I have carried out in the meantime.

I am pleased to say that our local Dr Malcolm Holding, (retired) was kind enough to supervise my actual testing of the turves water holding capabilities.

During the first week of March I decided to spend some of my time looking into the condition of topsoil on six various local farms ranging from an altitude of a few feet above sea level, and up to 1,800 ft for the highest, so by the tenth of March I had collected a one foot square, and six inches deep turve, off six various farms, I decided to give these turves the same conditions as they had experienced during the previous wet winter period, so on two applications they were given the equivalent of one inch of rain, this amount of water weighed 6lbs per square foot of area, equal to 54lbs a square yard, that works out at just under One hundred and Sixteen tons of water per acre.

We placed the turves on a slight average slope and measured the water that ran off the top, also the amount that penetrated down through to the subsoil or the rock, and by subtracting the sum total of these two we had the amount of water that the turves or soil was capable of absorbing if only for a short period of a few days.

The results were an eye opener, basing the amount of water as above, the figures came out as follows, all per ton of water per acre, giving us a fairly accurate but slightly rounded off figure,

Sample No 1.
Thirty eight and a half tons of water per acre ran off the top surface, and an equal amount ran through to ground, therefore exactly Thirty eight and a half tons of rain water was absorbed by the top six inches of soil.

Sample No 2.
Sixty five tons per acre ran off the top surface.
Forty tons ran through to ground.
Therefore only Fifteen tons was absorbed.

Sample No 3.
Forty tons ran off the top surface.
Seventy tons ran through to ground.
Therefore Ten tons was absorbed.

Sample No 4
Fifty tons ran off the top surface.
Fifty tons ran through.
Therefore twenty tons was absorbed.

Sample No 5.
Thirty tons ran off the top surface.
Fifty tons ran through to ground.
Therefore forty tons was absorbed.

Sample No 6.
Forty tons ran off the top.
Seventy tons ran through to ground.
Therefore Ten ton only was absorbed.


I would point out that none of the above are arable farms, and I would expect the amount of water run-off to be considerably more had this been the case, with the majority of the run off, from the top of the subsoil due to excessive panning, as opposed to running off the soft surface.

On completion of the above tests I then carried out the usual Jam jar test to assess the condition of the soil in each case, and found that the humus content was very low indeed in each sample, but No 1 and No 5 were by far the best for humus content, and therefore these were capable of absorbing more rain water as the above figures show.

The test that I used was as follows, I half-filled a jam jar with a sample of soil off each turve, this soil was then loosened up thoroughly and the jars filled to the top with clean water, they were given a good shake to mix them up after labelling, these were left to settle for a few days, stones and grit drops to the bottom, heavier soils follow and the amount of lighter humus settles on top, with some possibly even floating on top, very fine clay can take up to a week to settle. Afterwards these samples were left to dry naturally in a greenhouse, and it took twelve weeks before they were completely dry, each sample except No 1 and No 5 finished up a solid cake of soil, like a soft stone, whilst No 1 and 5 being what I think the best conditioned soils remained friable.

I put this down to the use of Artificial fertilizers over the years, which I think are made up of, and contain a small proportion of active ingredients, that are bulked up with cheap fillers, to make them possible for farmers to spread, these fillers I think clog up natural soils and causes rain water to run off the top rather than being absorbed and/or retained for a period of time at least, in the top soil.

These samples were taken during the second week of March 2014 this year, following an extremely mild and wet winter, the ground temperature was quite warm at the time, despite this I found no evidence of active Worms in any except No 1 and 5, significantly I found a few dead worms on the field of No 2 when I returned the turve three weeks later, this farmer had dressed this field lightly with well stirred slurry the previous day, what really shocked me was the fact that I also found a tight ball of young worms each one slightly over half an inch long, all intertwined and dead, obviously killed by exposure to the sun after coming up to the surface for air, when the stirred slurry was applied.

I cannot believe that I missed the opportunity to count these, and could truly kick myself for being again in a slight hurry, this ball of immature worms was a little larger than half the size of a golf ball, so contained quite a few, probably around fifty baby worms in number, to say the least. (I have seen mature worms forming a ball like this when I open my garden wormery, each one individual trying to escape the sun's rays).

I should mention here that in each case I took care to avoid areas that had been compacted by the usual farm traffic. Soil compaction could only cause much more immediate run off and I have completely ignored this very important aspect of the problem.

A few notes regarding each Sample in turn.

No1.

Taken off a fairly high altitude mixed Beef and Sheep farm of light loamy soil, this farmer regularly slits his grassland to improve soil aeration, and has for the last few years pumped his slurry pit out over his fields without stirring, all the solid effluent he then carts out onto a heap, and this is spread on his fields next season as a dry material, he finds that this dry matter and especially the unstirred dirty water does his fields a power of good, the little extra effort has proved very worthwhile, as his land is producing more and is steadily improving, whilst his monetary costs are steadily reducing.

No 2.

Again a Beef and Sheep farm, this land is very well farmed, this field has been cut for silage two or three times a year for quite a few years, and has been treated with stirred slurry consistently, mostly grazed by sheep each winter.

No 3.

A strong loam, this land was purchased by the present owners three years ago, previously it had been rented out for almost forty years, and probably only high nitrogenous fertilizers used to produce crops and grazing during this long period, the result being that due to the very poor amount of humus left in this ground it had a very low ability to hold rainwater, even after three years of careful husbandry and an annual light dressing of dry fibrous animal manure, this land was much improved at the time of testing compared to the condition it was in, when purchased, the new owners will need a few more years to restore the fertility to an acceptable level, after the abuse of modern farming whilst it was let out.

No 4.

A heavy loam, this is an amenity field that has not been farmed, again for 40 years or so, a crop of grass has been taken off once yearly and practically nothing else done, a thick thatch of dead grass on the surface caused more rainwater to run off the top, and the poor humus content of the soil allowed more water to run through to ground.

No 5.

A totally organic smallholding, a strong loamy soil, no artificial fertilizers used in living memory, less water ran off the top of this land than all the others indicating that artificial fertilizers do clog or affect the porousness of farm soils, there were more fibrous roots in this soil due to slightly more weeds, this land retained water equally as well as the No1 sample which was much the best farmed land of all the samples.

No 6.

A red sandstone soil. A rather intensive Dairy farm, this land I understand to be mostly grazed by milking cows, I noticed that the grass was rather sparse or even thin.

It is not for me to judge the growing conditions of this farm or any of the others, my concern is today's lack of blotting paper like ability of our farmland soils to absorb and delay the run off of heavy storm waters.

Although these samples all showed a very poor amount of humus, which is the essential material that holds water, our local rivers here in South Wales just about managed to cope with the equally excessive amount of rain we experienced in this area during the same winter period of early 2014.

We certainly cannot afford to let the poor condition of our farm soils are getting into today, to deteriorate much more.

The excessive flooding of the English rivers are mostly due to the recent modern methods of mono-farming, mainly triggered by our government's cheap food policies. Our farmers are the custodians of our precious lands, they should not be harassed so much that they need to abuse their soils just to make a living, and thereby upsetting our local climates.

I believe it is time to do some serious thinking to address this problem as a local one in our country, it is not doing any good at all blaming global warming, quite recently in a gardening programme the presenter asked our top and very best green-keeper what was the secret of keeping and growing good looking grassed areas, his immediate answer without hesitation was "Aeration, Aeration and more Aeration".

I believe if farmers could be helped to afford the time and the expense of aerating one third of their land each year, this would alleviate the flash flooding considerably, they would also realise very soon that this would greatly benefit their crops as well.

Natural rain water is the catalyst that enables the nutrients deposited in the sea to be recirculated back to benefit our soils, to do this rain must be able to penetrate our farm lands efficiently.

Water has always been a good servant to humanity, but if it is allowed to become our master, god help us all, the problem must be tackled at source, urgently before it becomes our complete master.

Yours with respect,

W John Davies.

P.S.

I emphasize that the contents of this last chapter is totally my interpretation of facts found while doing this work, as said it was a real eye opener to me personally.

I can only hope that what has been written will generate sufficient interest in the people that are in a position to do something about it, the flash floods are a new serious problem for those that have already suffered, the answer is relatively simple, but it needs to be taken in hand quickly, diligently and seriously.

Some months later, I received a response from DEFFRA, the Government department that is concerned with Agricultural affairs in our country, they said that they had the flooding situation within their sights, and they were on track to alleviate the problem by the year 2030, fifteen years hence!!

Ironically the word Deffra in the Welsh language means 'Wake Up'. Feeling rather disgusted with their high handed attitude, I rest my case, without further comment.

# Chapter 23

Growing potatoes, Past and present.

My earliest memories of growing potatoes was as a child of ten years old, all of our farmers were compelled to grow this crop due to the second World War, our local farmer asked me to be available on Saturday to help drive out animal manure on to the potato field, the potato rows had been opened the day before using a Double Tom, this was a double sided plough that opened a single deep row across the already cultivated field, this was pulled by two horses, that went back and fore until the required acreage had all been opened into drills.

On the Saturday, there were two men using hand forks, filling the carts, from the muck heap on the yard, I was now promoted, and my job was to drive the horse and the full cart out to the field, on arrival I then had to lead the horse along the row stopping every so often, another man would follow the cart and pull out enough manure into a heap each time I stopped, this man was using a fork that had its prongs bent at ninety degrees off the wooden handle, this was known as a Crook, when empty I would jump into the cart and drive the horse at a trot back into the yard, where by now another cart was full waiting for me to take it out again, this went on all day, or until the field was covered with dung heaps.

If there was daylight left then, it was all hands on deck to spread the heaps along the rows, the potatoes were then planted by hand into this dung, with a light sprinkling of some artificial fertilizer also sown by hand, carried in a Seed Lip, this was like a long bucket some 30 inches long and shaped to fit around your tummy, with a strap around your shoulder to carry it, once a few rows were planted a single horse would then be used with the Double Tom to close the rows up over the planted potatoes, this horse had to be very steady, as he had to walk on top of the ridge to avoid stepping on the potatoes.

The potato fields were regularly hoed by hand during the growing season to control the weeds, and the crop was usually lifted during an extra weeks school holidays called, "Potato week," when everybody able-bodied living in the area, were expected to help.

Gradually mechanisation has taken over, during the Fifties and Sixties, soon after the second world war large acreages of Potatoes were grown in Pembrokeshire mainly targeted towards the very early markets, as the county especially around the seaboard, was fairly free from frost, but by today (1914) the emphasis has now swung somewhat towards main crop growing, as handling facilities have improved, there is now a large local packing facility developed in Pembrokeshire to make marketing considerably easier all around.

I would imagine that heavy crops of mature potatoes should be more profitable, than lifting light crops to catch the early market as it used to be.

Since retiring I have enjoyed growing Potatoes in my garden, and I still grow enough for our household all the year round, and thankfully I have been extremely lucky to have enjoyed health good enough to be able to do so.

During the last decade I have endeavoured to grow all my potatoes as organically as possible, and also to Show standard, this takes a little more care and time, but I found it interesting and fulfilling to always try to do things a little bit better.

Each year I buy the best available seed from Scotland, they usually arrive early January unless the weather is too cold, I then wash these thoroughly, and discard any with blemishes that this washing shows up. They are then graded according to size, as large seed produce an earlier crop, small seed will produce equally well, but they need a few weeks longer to catch up.

Boxed up in single layers it is important not to leave them shiver, should the winter weather turn cold. I grow a few earlies in pots, and these are kept indoors until they start to come up through the soil, they are then put out where I can watch them, so that I can cover them with fleece, should the weather threaten frost, normally I have nice earlies around twelve weeks after planting. A bag full of these first little beauties are put in the freezer, and they are kept for our next Christmas dinner.

As soil conditions warm up I plant outside, a few are also planted in 23 litre plastic pots, filled only with fine Peat and some fish, blood and bone organic fertilizers, these are grown for showing , and are grown like this to avoid any blemishes on the finished skins.

These pots are then half buried in the soil to help them retain moisture, water is very important if you want the best out of your potatoes, if they are allowed to dry out and you water them heavily, they will drink it up too fast and they will develop cracks.

So if yours should get very dry, water them lightly twice a day until the soil gets nice and moist again, likewise if a storm threatens and you think your spuds are on the dry side give them a little water before the heavy rain arrives. When I plant the fresh shoots or sprouts are looked at, if I want large potatoes I thin out the sprouts or chimes, to two or three, for an average good crop I leave four sprouts, as normally the more sprouts you leave on, your new potatoes will be smaller in size, but more numerous.

If you leave a few more inches width between the rows and also between the seed along the row you are guaranteed to get a better crop, as recently we have realized that a better degree of light will definitely produce a better crop, it is also important to open your rows from north to south.

Should you decide to have a go at showing potatoes very good luck to you, to choose the best showing varieties you will find these marked in all the best seed catalogues.

The most recent development that we know to be successful, after using it for around three years, is this, when spraying to protect our crops from blight we have tried and are happy with fresh Milk.

One pint diluted with three pints of water, think about it, fresh milk is alkaline, as it turns sour it becomes acidic, it has to date done no harm to our crops. You do not need to spray for blight until the minimum temperature at night reaches 15 degrees centigrade, below this blight is not active.

I believe that it is very important to have very clean water to irrigate potatoes, personally I purify the water I use on my tomatoes, and potatoes, being the same family, with Potassium permanganate crystals dissolved in the water. The results, I have found, are much better. To finish I wonder if you can believe that a spray of caustic soda will not harm your potato haulms, we have used this to delay blight, but the milk one is much easies and safer to use.

If you grow your crop in the same plot for too long, blight may get into your soil, and that means trouble.

You will produce better crops of all vegetables including potatoes, if your soil is well soaked with water and three table spoons of Domestos added to each gallon, I use five gallons of this mixture to every square yard of my garden, this is done during a spell of dry weather in early winter, and I have done this every other year for a few years by now, this may seem strange, but I assure you, this is true and very effective.

I have named Domestos as this is what I have used at this dilution rate, resulting in much improved crops all round, you will notice that your worms will increase dramatically as well, and that must only be good.

Other brands may do as well, but I can only vouch for Domestos as this is what I have used and therefore know is safe to use.

It has been well established in my mind, that well stirred slurry kills worms in agricultural land, I feel sure this practice sooner or later will be stopped, probably farmers will stop voluntarily.

If so I can only wonder whether the last few paragraphs regards Domestos will become general practice on grassland?

This morning in early winter a friend of mine rang up concerned that his suckling cows still with calves on, were not showing any interest in the good quality silage he had put before them a few days previously, apparently, he had an agreement with his neighbour that he could keep his cattle on a building site, this being a field that had nothing done to it for some time, it was full of tall dead grass and weeds including large thistles also dead. He was shocked that his animals were completely happy browsing this stuff that seemed so useless. Are these cows that are looking extremely well, telling their owner and farmers in general that they don't need food too rich in proteins, the good silage is there for them but as yet they have completely ignored it.

In Chapter 5, I mentioned the academic that mixed roadside grasses in with his good hay, he said it was for the benefit of his few milking cows

I sit here, half tired and contented,
wishing extremely good luck to each and all
of my readers.